TEST YOUR CAR BUYING IQ

1. Should you buy during "year-end clearance" sales . . . invoice sales . . . or cashback incentives?

2. Should you purchase rustproofing?

3. Can you negotiate the price of an extended warranty?

4. Should you clean and repair a car before trading it in?

5. Are most foreign cars worth more than American makes in five years?

6. Are dealer rebates or reduced interest rate programs a good deal for the car buyer?

Answers: (1) NO   (2) NO   (3) YES
(4) YES   (5) NO   (6) NO

These are just some of the questions you'll have to answer in order to buy the very best—and save money, too. So, when you go shopping for a car, bring along an expert—

HOW TO BUY THE MOST CAR FOR THE LEAST MONEY

Negotiate the Lowest Possible Price for a New Car Every Time!

HOW TO BUY THE MOST
CAR
FOR THE
LEAST MONEY

Negotiate the Lowest Possible Price for a New Car Every Time!

DANIEL M. ANNECHINO

A SIGNET BOOK

SIGNET
Published by the Penguin Group
Penguin Books USA Inc., 375 Hudson Street,
New York, New York 10014, U.S.A.
Penguin Books Ltd, 27 Wrights Lane,
London W8 5TZ, England
Penguin Books Australia Ltd, Ringwood,
Victoria, Australia
Penguin Books Canada Ltd, 10 Alcorn Avenue,
Toronto, Ontario, Canada M4V 3B2
Penguin Books (N.Z.) Ltd, 182–190 Wairau Road,
Auckland 10, New Zealand

Penguin Books Ltd, Registered Offices:
Harmondsworth, Middlesex, England

First published by Signet, an imprint of Dutton Signet,
a division of Penguin Books USA Inc.

First Printing, September, 1993
10 9 8 7 6 5 4 3 2 1

 REGISTERED TRADEMARK—MARCA REGISTRADA

Printed in the United States of America

For my wife, Janet,
whose support and confidence in me
has made a dream become a reality.

Contents

Preface

I have spent over seventeen years of my life in the automobile business. My voyage started in April of 1975 and continues as I write this book. During that time I have held many different positions within the dealership structure. This has enabled me not only to fully understand the car business inside and out, but to analyze and dissect the minds of the car dealer *and* the consumer. During my tenure, I have worked in every position, from salesman to general manager. I have worked for dealerships that represent Chevrolet, Ford, Nissan, Mazda, Jeep/Eagle, Chrysler, Plymouth, Volvo, Audi, and Honda, and I've spent a period of specialization in trucks only.

Car buyers have always fascinated me. I am especially fascinated with the naïveté with which they attempt to outwit car dealers. Out of the fifteen million or so people who buy a new car every year, it is doubtful that even 10 percent get a satisfactory deal. And none of that 10 percent realizes a fair deal without the aggravation, anxiety, and trauma that have been bred into the car-buying process.

After a great deal of research, which included reading several how-to books for car buyers and personally interviewing hundreds of typical consumers, it occurred to me that car buyers, particularly women, are still being misled and may be in even bigger trouble than I thought. Please do not misunderstand me. Many of the books that have already been written disclose some sound information. But I find consistent problems with each of these books. This has motivated me to write a book specifically focused on women buyers (although men will find it equally enlightening). I am confident it will become the car buyer's bible.

Many of the books already written seem to focus on the concept that in order to achieve a good deal on a new car, one must enter the showroom with a suit of armor and engage in a bloody battle. Another main theme always seems to be how to achieve the lowest price on a new car. There certainly is nothing wrong with this concept. But you will learn from my book that there is a dramatic difference between the "best price" and the "best deal." Perhaps most important of all, even I, with my seventeen years of automotive experience, find these books confusing and hard to follow. If they are confusing to me, I can't help but wonder how the average consumer reacts to them. My book is designed to simplify the automobile-buying process and to arm you with vital information that will make your buying experience a rewarding one. There is a way to turn the battlefield into a chess tournament, if you use your brain and remain open minded. This book can shed a new light on car buying, car shopping, and the negotiating process.

Car dealers from coast to coast have had the dubious distinction of being among the most disliked and distrusted business people in the world. In some cases the reputation is deserved; in other cases it is unwarranted. All industries and professions have their black sheep, but *most* car dealers across the country are considered common thieves by the consumer. Why have car buyers categorically classified car dealers as dishonest, conniving, and deceitful?

The truth of the matter is that there are no more or no fewer black sheep in the car business than in *any other business*. The situation is amplified and distorted because the consumer has concluded that it is virtually impossible to get a fair deal when purchasing a new car. Thus, car dealers must all be crooks! This is simply not the case. It is the shortcomings and limitations of car buyers themselves that allow car dealers to dominate them. Let us just call it "survival of the fittest."

The car buyers of the world should stop pointing an accusing finger at car dealers, take a close look in the mirror, and face reality. There is only one reason why the elusive best price, a high trade-in allowance, competitive finance rates, and a good after-the-sale service deal continue to outdistance them. The American car buyer is simply no match for the shrewd, educated, and cunning car dealer. Car buyers should not incriminate car dealers just because salespeople are smart businesspeople.

Car dealers continue to get the best of car buyers because they outwit them at the negotiating table. They don't beat consumers because they break the law or lack basic ethics. They have consumers completely figured out, but consumers haven't a clue as to what makes dealers tick. Yes, there are crooks and dishonest car dealers out there. But the majority of them are hardworking, clever businesspeople who know the art of selling and capitalize on every opportunity to earn a profit.

This book will allow you to learn as much, if not more, about the car dealer than he has learned about you. As you read this book, you will recognize familiar situations that you have experienced in the past. You will understand the mysteries that have plagued you for years. The myths, the fallacies, and the gimmicks will be exposed and defined. The car-buying process will take on a new light; you will be the predator, the car dealer the prey. I *strongly suggest* that you read the book once, quickly, and then a second time, slowly. It is a truly eye-opening revelation. Read and be enlightened!

Introduction

For over fifty years, the car-buying process has plagued the American consumer. Whether you're actually buying a car or simply shopping around for one, the anguish remains the same. Survey after survey has proven with little doubt that every consumer shares one common concern: Nobody enjoys shopping for a new car and furthermore, nobody is *ever* satisfied that they get a fair deal.

Dozens of books and articles are written every year to help the consumer lessen the miserable task of car buying. Do they help? I don't think so and I can prove it by asking two simple questions. First of all, do you or anyone you know (ask your fellow workers and your aunts, uncles, and neighbors) ever *really* feel secure and confident when you shop for a new car? Secondly, when was the last time you walked away from buying a new car and felt *completely* satisfied with the price, trade-in allowance, interest rate, after-the-sale service, and overall feeling of contentment? If you are like most people, the answer to these two questions is obvious.

Why is it that even well-informed, educated consumers never seem to get a fair shake when they shop for a new car? Every minute of every day a car buyer armed with some new innovative buying system gets beat at the negotiating table by a cunning car dealer. What then is the answer? Should American consumers accept their fate and allow the car dealers to dominate them for another fifty years? Don't despair. Help is on the way.

The car dealer, general manager, sales manager, and salesperson of today are not the barbarians of yesterday. In today's ferociously competitive automotive marketplace, only those dealers who are well informed, educated, and completely in tune with the times will survive. A revolution has taken place. Not only have car dealers recognized that consumers are more informed than they used to be, but the old way of selling cars—the "slam, bam, thank you ma'am" system that intimidated the consumer for fifty years—simply doesn't work anymore.

Ironically, car buyers are faced with an even greater challenge—the car dealer of the nineties. Today's automobile dealer is the product of an incredible evolution. Necessity has forced him to abandon the crude selling tactics of the past. Car dealers today are magicians, masters of illusion. They have met the challenge of today's cautious shopper. They have outmaneuvered the careful eye of the attorney general's consumer protection department and they have risen above and beyond anything that nature has ever created. Yes, my fellow consumers, the savvy automobile dealer today can tell you to go to hell and have you look forward to the trip!

When you walk into the average car dealer's showroom today, you won't find a dozen five-four, bald, overweight, shabbily dressed, cigar-smoking old-timers who greet you by exclaiming "Can I help ya?" What you will find is a professional salesperson who has a demeanor that is as smooth as silk. He will be well dressed and will answer virtually any question that you might pose about any car or option available at that particular dealership. He will be able to tell you why the cars he sells are superior to those of the competition. He will take you on a pleasant and informative demonstration ride and he will do a complete "walkaround" presentation of the car you are interested in. When you enter the salesperson's office he won't pressure you into making a hasty decision. He *will* do everything possible to consummate a deal today, but he won't beat you into submission. He'll be polite and he will display a sincerity that will make you want to take him home with you and adopt him! But beware. What you see is *not* what you get.

The car dealers of today have survived extinction. After years and years of making their showrooms a battleground, they have finally come to realize that you simply can catch more bees with honey than you can with vinegar. And so the car dealer of the nineties has been born. He is cunning, educated, and informed, with a complete and in-depth understanding of the enemy—you, the consumer.

The car dealer of the nineties is more dangerous than dealers of the past. He is the perfect predator. He will invite you into his spider web, disguised as a showroom, and no matter what type of self-preservation method you use—a suit of armor, a buying system, or coaching from an ex-car salesman—you will lose the war. The most brilliant scientist, the greatest salesman, even the chief financial officer of a major corporation, simply is no match for the car dealer of today. The car dealer has adopted a new strategy: *To know thy enemy is to conquer thy enemy.*

The car dealer of the nineties is to be commended for his brilliance. He knows everything there is to know about you. He understands your buying habits, your instincts, what you like and what you dislike. He knows what excites you and what turns you off. He studies demographic profiles on you and he analyzes your psychographic makeup. He knows what motivates you to buy and he knows how to push your "hot button." But the one thing he has learned and mastered, which enables him to continue with his fifty-year dominance over you, is this: If he can convince you, through professionalism and a sincere interest in your needs, that he truly cares about you and desires to keep you a satisfied customer, then and only then can he really be a winner. Because now he has not only sold you a car for an exceptional profit, but he has kept you happy in the process.

The car dealer of the nineties will totally engulf

you with professionalism and accommodation. He will satisfy your every whim. But beware. As you drive your new car out of his driveway, bragging to your husband or friend that this is truly the most pleasant buying experience of your life, more than likely you just paid hundreds of dollars more than necessary. Little did you know that the well-dressed, soft-spoken, low-key salesman to whom you will probably send your mother when she is ready for a new car, just made an outstanding commission. That great deal you just got on the car of your dreams cost you hundreds of dollars more than it should have!

Not only did you pay too much for the new car, not only did you get considerably less than your trade-in was *really* worth, but you were charged a premium for the extras you added to your car. And you paid a much higher interest rate by using the financing that he assured you was competitive with local banks. Alas, the plot thickens! The evolution of the car dealer of the nineties has helped to carve another notch into his six-shooter and your checkbook is the victim. You are so overwhelmed with the illusion of victory that year after year you will return to his showroom for the same beating.

This book, for the first time ever, will give you the ability to completely overwhelm any car dealer, on any day, at any time. I will cover every area of the car-buying process. I will remove forever the misconceptions and fallacies that have plagued the car buyer time and time again. Step by step, I will show you how to accomplish what no consumer has ever accomplished. I will completely dissect the car dealer of the nineties and expose his weaknesses, his strategies, and his vulnerabilities. I will reveal methods of shopping for a new car that will allow you to control any situation. You will learn the difference between the best price and the best deal. I will specifically outline methods of planning and negotiating that will save you time and money. You will be able to outwit the most cunning salesperson in the world. *To know thy enemy is to conquer thy enemy.*

I will show you how to buy *any* new car for the lowest possible price. You will get the most for your trade, the lowest finance rates, and the best service after the sale. I will give you foolproof tips on recognizing deceptive advertising and help you avoid chasing deals that are too good to be true. Whether you are interested in buying or leasing, trade or no trade, import or domestic, you will be able to overcome every scam and gimmick, and every smooth-talking salesperson you encounter. Nobody will ever pressure you into buying a car again. No one will ever take advantage of you again. You will never pay for an option you don't want on your new car, and the overall buying experience will become a rewarding, fulfilling adventure. The car dealer of the nineties will meet his match: the car *buyer* of the nineties!

Why Women Face a Greater Challenge

During the outline, planning, and writing of this book, I vacillated daily on whether or not it should be specifically devoted to the female buyer. My dilemma was to be sure that women would welcome this book in the spirit in which it was written and not conclude that I am just another male chauvinist. My candidness throughout this book can easily be misinterpreted. I assure you that my intentions are honorable and if some of the truths I expose are offensive, then the creation of this book was in fact a prudent one.

My purpose and motivation for writing this book is twofold. First of all, I hope to eradicate forever the adversarial relationship between automobile dealers and consumers, particularly female consumers. Secondly, I would like to remove the unfair misconception that all car dealers are unscrupulous. Many times there is a fine line between making a profit and taking advantage of a customer. Hopefully, I can effectively define the difference.

Unlike any other workplace, car dealers' showrooms have defied the laws of probability. I cannot quote exact statistics, but an educated guess would be that only about 5 percent of all salespeople in car showrooms across the country are women. In contrast to this, every other commission sales industry has seen the male-to-female ratio lean toward women.

For example, the real estate industry, advertising sales (radio, television, and print), furniture, appliances—just about every commissioned sales position you can think of—has seen the female population increase dramatically. Why has the automobile dealer's showroom been impervious to this trend? A showroom is the epitome of male dominance. Car salesmen view a female salesperson as an invader and unjustly feel that the showroom is the only haven left where men can be men.

They will do everything humanly possible (and some inhumane things!) to squeeze a woman salesperson so hard that she just can't cope with the pressure and will ultimately resign. Sexual harassment, degradation, sarcasm, vulgarity, and isolation all seem to peak the moment a woman is hired as a salesperson. And I truly believe that many times this unfair treatment is almost involuntary. It is the nature of the beast. To effect change would be like trying to teach a cat not to chase birds and kill them. I am not endorsing this behavior; I'm only telling it like it is.

Only a handful of women across the country

have the fortitude, stamina, and desire to over-come the sacred domain and territorial selfishness of car salesmen. During my seventeen-year tenure in the automobile business, I have seen woman after woman enter the showroom with the confidence of Mohammed Ali entering the ring, and a couple of months later throw in the towel. It is not fair. It is immoral and unethical. But it is a reality that exists and I don't know if it will ever change.

How does this relate to women car buyers? If you think that it is difficult for a woman to work side by side with a car salesman, then it is truly mind boggling to imagine how most car salesmen perceive women car buyers. When a woman ventures out to purchase a car without the assistance of a man, she is considered to be the easiest of all prey. Even the most honest, ethical salesman in the world can't resist concluding that a female car buyer is the quickest route to a *big, fat* commission. And he usually accomplishes his mission.

You can be the shrewdest, toughest, most cunning car buyer in the world, but the minute you enter a car dealer's showroom alone, you may be in for a very insulting and exasperating experience. Now I am not suggesting that every woman in the world who has ever purchased a car on her own has been treated like a second-class citizen, but most women will agree that buying a car alone is not an easy chore. Even if a salesman isn't insulting, or a sexist, he is almost certainly patronizing and condescending, and many times this action is totally innocent and unintentional. As I said earlier, it's simply the nature of the beast. You not only have to deal with the many challenges of buying a car, but unfortunately, nature has added one more little curveball to the game.

And to complicate things even more, I have even worse news for you! Over the last few years as a general manager, I paid very close attention to the growing number of female car buyers. Over 50 percent of our sales resulted from female buyers who made the car-buying decision completely on their own. I monitored our sales figures very carefully. We consistently made more profit with women buyers than with men.

Why? There are two reasons. First of all, a woman is much more cautious and suspicious of a salesman when first encountering him. However, once she is convinced that he is sincere and ethical, she becomes much more at ease and trusting than the average male buyer. Once a deal is agreed on, she lets her guard down and assumes that everything is on the up and up. A costly mistake. More often than not, the salesman easily talks her into buying dealer-installed accessories, extended service agreements, rustproofing, paint sealant, fabric protection, and dealer financing. The end result is a happy customer and a total deal that yields an above-average profit. It would be difficult to accuse the car dealer of anything except being a smart businessman.

Secondly, a woman who feels insecure about purchasing a car on her own, but buys one alone anyway, will usually ask a male friend for advice—another tragic mistake. The typical male buyer is an egomaniac when it comes to purchasing a car. He believes that he has all the answers and that he can easily outwit any hotshot salesman. To put it bluntly, you can take the advice that he gives a woman on how to buy a car and seventy-five cents, and buy a cup of coffee. His coaching is all but worthless. Men have not fared well in the car-buying process either. They've done a *little* better than women, but nothing to brag about. Asking an uninformed man how to buy a car would be like asking the big bad wolf if Little Red Riding Hood should be left home alone.

So how does a woman deal with the unique problem that nature has added to the list of car-buying obstacles? Well, I've evaluated and reevaluated the situation in my mind, and I've done some research. In fact, I've had many intense conversations with a number of my closest female friends and business associates. Most of these women are well educated, career oriented, intelligent, and very independent. I've quizzed them and asked them some in-depth questions about their car-buying experiences and their overall strategies. I am staggered by their naïveté and their ignorance of the car-buying process. They are far from being stupid. But they are *totally uninformed* and *completely misinformed*.

The *only* thing that the female buyer needs to protect herself from all of the pitfalls and misconceptions associated with the car-*shopping* and car-*buying* process is reliable, accurate information and a game plan to apply this information. I am absolutely confident that this book will accomplish this mission precisely.

If you learn from this book and employ the methods of negotiating that I suggest, avoiding the pitfalls, fallacies, and misconceptions, I assure you that the car-buying process will take on a new light. You will enjoy a new confidence, even if you are a relatively timid person. When you enter the car dealer's showroom, the information unveiled in this book will benefit you time and time again. I will teach you how to handle any and all car-buying situations, no matter how difficult they have been in the past.

1

The "Best-Price" Pitfall

There is no comparable mistake to the catastrophe that results from a car buyer shopping around for the best price. It is without a doubt the most disastrous thing that a car buyer can do. The "best-price" pitfall is the car buyer's worst enemy and the car dealer's biggest ally.

Although each and every one of us is unique, with different emotions and ego drives, the vast majority of car buyers fall to a common trap when shopping for a new car—the best-price pitfall. Whether the individual goal is to achieve the best price, the highest discount or trade-in allowance, or the lowest monthly payment, the method used is the same. After less than fifteen minutes of entering a showroom, most car buyers have their hands around a salesman's throat demanding to know what the best price on a certain car is, or how much their trade-in is worth or what their monthly payment will be.

If I were a gambling man I would wager a fair

amount of money that *every* consumer who has *ever* shopped for a new car has asked a salesperson for his best price. Sounds reasonable enough. After all, it is the goal of every car buyer in the world to achieve the lowest possible price when shopping for a new car. It only makes sense to ask for it, right? Wrong. Dead wrong.

Perhaps one of the most difficult things for the consumer to accept is this paradox. The *only* possible way for a car buyer to ensure the best price on a new car is to never—I repeat, *never*—ask for it. I realize that this doesn't make much sense right now, but I assure you, my explanation will shed a new light on the best-price pitfall.

There are three reasons why a car buyer should not ask a salesperson for the best price. I will cover each of them in detail so that the importance of each is clearly understood.

First of all, in most cases, a typical salesperson cannot give you a best price because he probably doesn't know *exactly* what it is. Yes, it's true. Whether you like it or not, buying an automobile is the purest form of horse trading that has ever existed. The sales manager's job is to sell every car on his lot for sticker price, or as close to sticker price as possible. In contrast, it is your goal to buy that car for as low a price as the manager will accept. But what *is* the best price? The best price changes with each deal, with every customer, and with each passing hour!

There are many factors that affect the ultimate price that one pays for a new car. Some of the key factors are the size of a dealer's inventory, the time of the year, the time of the month, and supply and demand. Oh yes, supply and demand. Remember when Hondas were selling for $1,500 to $2,000 *over* factory sticker? With stable demand for Hondas and plentiful inventories, you can now buy a new Honda for about the same percentage of markup as a Chevrolet. But the most significant factor that determines the price of a new car is the ability of the individual consumer to outnegotiate the car dealer. Let me repeat this crucial revelation: *The best price on a new car is almost solely determined by the negotiating ability of the individual consumer.*

Some time back in 1925, a car buyer walked into a Ford dealership and admired a beautiful new Model T. He glanced at the price of $850 and took a deep breath. When a salesperson approached him with the usual "Can I help ya?", the car buyer replied, "If you can sell me this car for $800, I'll drive it home today!" The very next day, a different customer said, "If you can sell me this car for $750, I'll drive it home today." Although this is oversimplified, it explains how the whole horse-trading part of car buying started. Since that moment in time, each and every best price that has ever been given on a new car was the result of one's horse-trading ability.

Many people will argue that buying a new car should be like buying a loaf of bread. It should not be a horse-trading game at all. But remember that it is the consumer, not the dealer, who benefits from the horse-trading business. The hard, cold truth is that if you truly do not want to horse trade, then simply pay the dealer the sticker price and that will be that, just like the loaf of bread.

See what I mean? The sooner the consumer realizes that horse trading plays a vital part in their ability to achieve a fair price on a new car, the better off they will be.

For over fifty years, hardheaded consumers have fought tooth and nail to avoid the horse-trading arena, viewing it as an aggravation and an inconvenience. You will soon come to realize that horse trading is, in fact, the consumer's greatest weapon. Once the process is put into proper perspective, it will become a friend, and using it as a tool will make car buying a joy.

So what have we learned? The salesperson cannot give you the best price because it is a number that can only be negotiated. It is worth mentioning at this point that even if salespeople *did* give legitimate best prices, what value would these best prices have, anyway? Isn't your ultimate goal to buy a new car for what you consider to be *your* best price? Who cares what the *salesperson's* best price is? Your main goal is to achieve *your* lowest price, not the dealer's.

The second reason you should not ask for the best price is this. Salespeople quote prices every day. Just because they haven't a clue what the sales manager will or will not accept doesn't stop them. They will throw prices at you day in and day out. The problem is that the prices they quote you are all but worthless. Am I insinuating that all salespeople are liars? Absolutely not. But they are survivors.

Put yourself in the salesperson's shoes for just a moment. When a customer asks a salesperson for the best price on a new car, what the salesperson *hears* is this: "Would you give me *your* best price on a new car so that I can take that best price and use it as a hammer to get the dealer down the street to give me an even *lower* best price?" Think about it for a moment. Isn't your *real* mission, when asking for a best price, to find that even lower best price?

A salesperson realizes that no sane consumer would walk into a car dealership, ask for a price, and buy for that exact price that day. It just doesn't work that way. All car buyers shop around for best prices before they make a buying decision. Salespeople know that sooner or later the car shopper is going to get really annoyed with the shopping game and return to the salesperson who gave him the lowest price. Salespeople also know that if the best price they give you is higher than the one you get from the dealer down the street, you probably won't come back a second time. Thus, the lowball price is born.

The scenario goes something like this. Ms. Jones spends a couple of weeks gathering price information from four different car dealers. When she is satisfied that her lowest price is a legitimate one, she goes back to the dealer who gave her the lowest price, prepared to buy immediately. When the salesperson writes the deal up and takes it into the sales manager for an approval, he returns with the classic "break the lowball" story. "Gee, Ms. Jones, the other day when you were in, I was *sure* I could have sold you the car for $11,000, just like I quoted you, but my boss says that we only have one of those hard-to-find little babies left in stock. I *really* could have sold you the car for $11,000, but today we need a little help." Sound familiar? This classic lowball will get you back to the dealer a second time, every time!

The bottom line is this: Every salesperson at every car dealer across the country knows that to give a customer a *real* best price (whatever that may be) is to never see them again. It is an occupational hazard that salespeople face with every best-price shopper. As a result, most salespeople try to be vague, but when a consumer insists on getting a best price they usually give them a lowball price that assures the salesperson that the customer will be back.

The situation is simple. As long as car buyers continue to shop around searching for the elusive best price, the end result will always be the same. Frustration, aggravation, and the perpetuation of the lowball. They will never find the best price by asking for it—not fifty years ago and not fifty years from now.

The third reason the car buyer should never ask for the best price may be the most important reason of all. We have already established the fact that all car buyers shop for the best price. Consequently, all car dealers focus their strategies on the best-price shopper. Most salespeople have had extensive training on various tactics necessary to outmaneuver the best-price shopper.

Salespeople *expect* you to ask for the best price and they are prepared to outwit you when you do. Sales meeting after sales meeting is devoted to training salespeople in various techniques to successfully handle the best-price shopper. Believe me, they've got your number and are trained to overcome even the shrewdest best-price shopper. Nothing you can say or do will be anything new and innovative to the seasoned salesperson.

Now what do you think happens when a car buyer walks into a car dealership, spends an hour or so gathering information, and leaves without *ever* asking for a price? The salesperson is taken completely off guard. He becomes a little paranoid and very confused. Car buyers must realize that when it comes to buying a car, control is everything. By avoiding the best-price pitfall, you have achieved a strong position of control over the salesperson. When you ultimately begin the negotiating process, you will be in an advantageous position. Salespeople, sales managers, and dealers themselves haven't even begun to address the challenges faced when encountering an unpredictable consumer who isn't on a best-price mission. You see, car dealers think that they've got you pegged, but when you don't react the way they expect you to, it throws them into a tailspin. Control *is* everything. If the car buyer wishes to end up on the winning side of the negotiating table, then she must establish control the minute she comes face to face with the salesperson.

As you continue reading the chapters that follow, you will begin to understand that best-price shopping is an exercise in futility. It is primitive and dangerous. You will also understand how the overall strategy of the wise, well-informed car buyer will not include *any* price shopping.

2

Timing: An Important Piece of the Puzzle

Have you ever heard the cliche, "timing is everything"? When it comes to buying a new car, the *right* timing can save you hundreds of dollars. Likewise, the *wrong* timing can result in a very expensive mistake. Let us analyze the reasons.

Selling automobiles is a numbers game for the car dealer. As we discussed in Chapter One, every deal on every car is negotiated. Therefore, the individual price that one buyer pays may vary considerably from what someone else might pay. A car dealer does not make the same profit on every car sold. It is a game of numbers, but it is also a game of *averages*.

Some deals, particularly those made by naive consumers, can make a car dealer as much as $2,000 or $3,000. In fact, on many high-demand expensive cars—those over $35,000—a dealer's profit can be even more. Automobile manufacturers put a tremendous amount of pressure on car dealers to maintain aggressive sales objectives. Consequently, the dealers pound away at their sales managers to meet these objectives.

Many times a car dealer who does not meet factory objectives will lose incentive money or allocations of specific model cars that are the most desirable. This can be a very costly mistake. A car dealer will do everything possible, including selling a car for a fifty-cent profit, *if* he needs the sale badly enough. This is why timing is so important to the wise car shopper. If you buy your car at just the right time, you will save a bundle.

Automobile dealers measure the performance of a dealership based on a term called *average gross profit*. Each department within a dealership is a profit center and must survive on its own merit. Typically, there are six very defined departments within a dealership: new car sales, used car sales, finance and insurance, service, parts, and a body or collision repair shop. The area of concern for you, at least at this point, is to understand the numbers game in the new car department.

Volume is the key and average gross profit is the buzzword. One fact remains consistent from dealer to dealer across the country. On some deals, the dealer makes a fortune, on others he barely makes a dime. Your goal is to strategically plan your purchase at the optimum time to ensure the best possible deal. Let the uninformed consumer pick up the tab for the car dealer. You can be one of the few, well-informed car buyers who contributes to the all-important volume number but *not* to the profit number.

No matter how much a car is in demand, no

matter what song and dance the salesperson gives you, the smart, well-informed buyer can always be on the winning side of the negotiating table *if* the timing is right. It doesn't matter who the dealer is or where he is located. The best deal can be achieved at any dealer you choose, on any car you choose. As long as you fully understand the difference between average gross profit and volume, you will be able to capitalize on timing. Now let's get inside the brain of the car dealer to fully understand the volume and average gross profit philosophy.

I am going to use some arbitrary numbers just as a point of reference. The actual numbers will vary considerably from dealer to dealer. The principle will remain the same regardless of the variation in these numbers.

We will assume that a particular dealer analyzes the expense structure of his dealership and concludes that the new car department must generate $120,000 gross profit per month to earn a fair profit. This can mean several things. He can sell 100 cars at an average gross profit of $1,200 each. Or he can sell 50 cars at an average gross profit of $2,400 each, or any combination thereof. Bear in mind that *volume* is the key in today's ferociously competitive marketplace. A car dealer *has* to maintain a high volume of cars sold.

If the dealer's volume goal is 100 cars per month, then he must maintain an average gross profit of $1,200 per new car sold. Remember that the nature of the negotiating process means that the weak, uninformed customers will pay as much as $2,500 profit to the dealer, while the well-informed, educated buyer may buy the same car and contribute only $50 profit to the dealer. The only two things that matter to the dealer are to achieve his gross profit of $120,000 and to sell 100 cars. He couldn't care less how his sales manager gets the job done as long as the numbers work.

So when *is* the right time to buy a car? The answer to that question has several parts. Chapter Three—Tent Sales, Invoice Sales, Year-End Clearances, Etc., Etc.—clearly explains when *not* to buy a car. The best time to buy a car is when you are *fully prepared* to outmaneuver the most cunning salesperson in the world. And that is an enormous challenge. It will take a great deal of planning, number crunching, and research before you will be prepared to meet the car dealer of the nineties head on.

Achieving an exceptional deal on a new car is not an easy task. You must do your homework and you must also gather information during the shopping process (which we will discuss in a later chapter). Once all of this vital information is in the hopper, you will be prepared to enter the negotiating arena with total confidence. Remember that the right timing doesn't just relate to the calendar. Timing also means that you have assembled the required information and formulated a foolproof, flawless strategy.

Timing is capitalizing on the costly mistakes that the uninformed consumers continue to make. As cynical as it may sound, the majority of the car buyers in this country will *never* get a good deal on a new car. The average consumer will continue to try to outwit car dealers with prehistoric meth-

ods. Their loss is your gain. As long as car buyers
are content to pay car dealers hundreds of dollars
more than they have to for new cars, the average
gross profit theory will remain viable. If your timing is right and your approach skillful, the car
you purchase will contribute to the all-important
volume number, but little to the average gross
profit number.

3

Tent Sales, Invoice Sales, Year-End Clearances, Etc., Etc.

Most people who make a decision to purchase a new car start leafing through the local newspaper to find out which dealers are selling their cars at sale prices. When the buying juices start flowing, one thing is certain: The consumer wants a new car immediately and also wants to buy the car on sale. And where do you get the lowest price on a new car? Naturally, you check out the sales.

Some dealers advertise low prices, others rebates or reduced interest rates, and still others include "free" options with the purchase of a new car. As you start to pay very close attention to radio, television, and newspaper ads, your goal becomes very confusing. It is a very difficult task comparing one ad to another and trying to determine where the really good deals are.

Every dealer claims to have the lowest prices, the biggest selection, the best service after the sale, the highest trade-in allowances, and on and on and on. How can you possibly distinguish the *real* sales from the come-ons? Let me put your mind at ease once and for all. All sales (with very few exceptions) are designed to do two things: to bring eager buyers in the front door of the showroom and to sell them a car for a fair profit. Period.

Yes, there are legitimate sales, but they are so few and far between that you would be doing yourself a real favor by avoiding them altogether. It simply is not worth the time and aggravation to sort through every sale and find the real one. Forget it. Besides, you don't *need* a sale in order to get a rock-bottom deal on a new car. All that is required is a little coaching and a lot of common sense.

Let us take a close look at the reasons you should avoid sales and special events. First of all, my personal experience, along with that of many other car dealers I have spoken to during dozens of sale events, has strongly indicated that car dealers make more profit per car during a sale than they do at any other time. Why? Because many times when a sales event is taking place, the car dealer's showroom is full of excitement. It is a virtual circus atmosphere and people get caught up in the hype. They get so wound up in the activity around them that they conclude that all these people wouldn't be here unless everybody was getting good deals.

Secondly, even though buying a car should be a well-planned, carefully executed strategy, many times people make impulsive decisions that they ultimately regret. Have *you* ever casually responded to a car dealer's sale with absolutely no intention of buying a car, only to find yourself driving one out of the dealer's driveway the next day? If you haven't, at least a hundred people you know have. It simply is a fact of life.

During sales events, dealers pull out all the stops. The refined, educated car dealer of the nineties temporarily becomes a barbarian again. Rest assured that when a car dealer spends big bucks on an advertising promotion, his sales managers and salespeople better use *every* tactic and resource available to close every deal possible. This is truly a no-holds-barred situation. The dealer may even risk jeopardizing customer satisfaction. He only cares about results.

The pressure exerted on car buyers during a sales event is so intense that many times a dealer will *sell* as many as fifty cars during a sale, but only actually *deliver* twenty to thirty of them. A strange thing happens to people when they buy a car impulsively. About an hour after the sale, they get a syndrome called Buyer's Remorse. They start to hyperventilate and question whether or not they *really* got such a good deal. This is because a buying decision was made so hastily that they second guess the validity of the sale price they received. And what do they do? The worst thing possible.

The next day (after spending the most restless night of their lives) they pick up the telephone and start calling other car dealers to try to tactfully find out just how good a deal they got. A tragic mistake. You want to talk about mind games.

Most salespeople can easily distinguish a legitimate telephone prospect from someone trying to
be sure that they got a good deal at a competitor's. Needless to say, after about thirty seconds
on the phone, the inquiry becomes totally trans-
parent and the shrewd salesperson capitalizes on
the situation by giving the frantic car buyer misin-
formation. By the time the conversation is over,
naive car buyers are convinced that they paid two
or three thousand dollars too much for the car
they bought on sale. Their palms get sweaty and
they get that knot in the back of their neck. Now
what do they do?

Many car buyers get so caught up in Buyer's
Remorse that they decide to cancel the deal they
made. That's when things can get really ugly. Do
you think for one minute that a car dealer is going
to let you off the hook and politely let you back
out of a deal without a fight? The dealer refuses
to return your deposit. You threaten to contact
the attorney general's office.

Who needs this kind of aggravation? The best
way to avoid Buyer's Remorse is to avoid sale
events. Buying a new car is serious business.
There is no room for impulsive decisions. You
cannot enter the negotiating arena until you are
completely prepared.

Here is another reason to avoid sale events. Did
you ever really take a close look at an average car
dealer's newspaper ad? Did you ever read the fine
print right next to the little asterisk? More often
than not, newspaper ads can be confusing and
sometimes deceptive. The sale price is in big bold
type, but the explanation of the sale price is writ-

ten in the smallest print possible. There is no such thing as a free lunch. More often than not, when you respond to a sale event, you will be a victim of a bait-and-switch ploy (a method of advertising a low price on a car and then switching the prospect to a more expensive car), or of some other kind of misleading advertising. Forget it. Do not, under any circumstances, respond to a sale.

About four years ago, the New York State Attorney General launched an aggressive attack on the deceptive advertising practices of some car dealers. He released a seventeen-page document that outlined very specific guidelines for dealers. To make a long story short, dealers were on their best behavior for about six months. Needless to say, the state of advertising slowly and gradually returned to normal. I can assure you that on any given day, in any average city, regardless of how astute the buyer, car dealers' advertising will be laced with vagueness and confusing claims.

Now that you know when not to buy a car, when *do* you buy one? As I mentioned in Chapter Two, Timing: An Important Piece of the Puzzle, you will not be prepared to buy a car until you completely assemble a plan that includes a lot of number crunching and research. However, the calendar is also of the utmost importance. In Chapter Two we also talked about average gross profit and the importance of sales volume to the dealer. With the knowledge that dealers, sales managers, and salespeople have specific volume objectives and, many times, cash incentives for reaching these objectives, the pressure exerted on them in the last five days of the month is at its peak. Manufacturers as well as car dealers project their sales volume objectives on a monthly basis. Most salespeople and sales managers can earn substantial month-end bonuses *if* they meet their particular objectives. What do you think happens when it's the end of the month and a sales manager or salesperson is three or four cars away from earning a big bonus? That's right. They will do anything to sell a car, including almost giving one away. Remember that it is average gross profit and volume that concerns the dealer. If you play your cards right and do your homework before you enter the negotiating arena, you will be able to capitalize on the end-of-the-month squeeze and ensure a rock-bottom price on the new car of your choice. And believe it or not, everybody wins. You get a great price on a new car, a salesperson earns a commission, and the car dealer helps to reach his objective.

When is the best time to buy a car? The last five days of the month. Now bear in mind that a sales manager's motivation can only be capitalized on if you actually take delivery of your new car *before* the end of the month. You will have to be sure that you give the dealer a couple of days to prep the car, do the paperwork, and install whatever options you may have purchased.

It is crucial for you to understand that this end-of-the-month strategy all by itself will not assure that you're going to get a really great deal. It is just one piece of a systematic approach. As you continue reading this book, things will really start to come together and become clearer and clearer.

4

Misconceptions, Myths, and Fairy Tales

The first three chapters of this book have convinced you, I hope, that long before you can enter the negotiating arena, you must first assemble a strategic plan. This plan will consist of research, number crunching, and a lot of homework. It is also necessary for me to devote one chapter to removing forever the old wives' tales that have been around as long as automobiles themselves. During the more than seventeen years that I have spent in the automobile business, I have found it amazing that so many people could be misled by so many misconceptions about the buying process. I strongly suspect that these deadly misconceptions and myths get handed down from generation to generation.

You know the story. An eighteen-year-old boy is about ready to purchase his first car and dear old dad, the veteran buyer that he is, sits down with Sonny and gives him his list of do's and don'ts. Of course this list was handed down from

Grandpa, who was the negotiator of all negotiators! Forgive my sarcasm, but the point is this: Everyone is an expert when it comes to buying a new car, but the truth is that few truly *are* experts. Even well-informed, educated, street-smart consumers fall prey to a number of undying fallacies.

Each of the following explanations will enable you to clearly understand just how susceptible and misinformed car buyers really are. Please bear in mind that many of the misconceptions I will try to illustrate are a result of real-life experiences during my many years in the car business. If you carefully review your past buying habits, I believe that you will recognize mistakes that you have made during the car-buying process.

All too often, car buyers become victims of the adage, "Can't see the forest for the trees." Consumers become entranced and obsessed with achieving the lowest possible price on a new car. Sometimes, this obsession makes them forget about other important components of the buying process that have a more dramatic effect on the overall cost than the price itself.

What is the benefit of buying a car for $100 over invoice if you purchase accessories for your car at inflated prices? Or what if you get an excellent price but the monthly payments are really taxing your budget? Or what if the interest rate is 4 percent or 5 percent higher than it should be? Suppose you purchase a car for an incredibly low price but find that after the warranty runs out, the car is in the repair shop every other week? And how about your trade-in value? Did the dealer give you a fair trade-in allowance? Lastly, is it worth it to buy a new car that has a very attractive price (because of a rebate or reduced interest rate) at the time of purchase, but depreciates to 30 percent of its original value after three years? If you focus your attention on price and only price, you may find that a seemingly good deal was not a good deal at all.

All of these factors are significant in determining what the real best price is on a new car. The overall ownership experience, from the day you pick up your new car until the day you trade it in, is the only viable barometer of a good deal. Following is some enlightening information on various topics that can dramatically affect the true measure of a good deal. They should be carefully considered *before* negotiating a deal on a new car.

Rustproof, Paint Sealant, and Fabric Guard

I can simplify any illusions you may have concerning protection packages available to preserve your car's life by saying one thing. They are almost completely worthless. Period. I don't care what chemical engineer you talk to or what your cousin's great-uncle told you, *all* rustproof, paint sealants, and fabric guards are as close to being worthless as anything can be.

The only real value of these protection packages is dealer profit. Typical dealer cost on these items is about $125 to $150—for *all three* items! Some dealers charge as much as $1,000 for these protection packages. There is absolutely no question that selling these after-market options often can be more profitable to dealers than the actual car sale. Stay away from them.

I believe that all automobile manufacturers have become very sensitive to and conscious of the overall quality of their products. That is why warranties have been extended beyond 12 months/12,000 miles. Sheet-metal warranties on most cars have been extended to at least 36 months/36,000 miles. And most paint finishes are guaranteed for about five years or more. The superior sheet metal, enamel paints with clear coats, the rust-inhibiting primers, and high-quality fabrics used in today's automobiles mean that prudent buyers can avoid all after-market protection packages. Don't waste your hard-earned money.

But even if cars were not built any better than they were a few years ago, the fact remains that the protection packages sold by dealers have negligible value. Many dealers add these options to their cars as soon as they receive them from the manufacturer. The logic here is additional dealer markup or additional dealer profit. Assuming that dealer cost on these items is $150 and retail is $699, even if they give the entire markup away in the form of a higher discount or trade-in allowance, it will create the dangerous illusion that you are getting a better deal than you really are.

Let me tell you specifically why rustproof, paint sealant, and fabric guard are all but worthless. First of all, from a scientific perspective, it is arguable whether or not the rustproof coatings do in fact inhibit rust. But assuming that they do, the method of application becomes critical. It has been my experience that all rustproof technicians (those who apply the rustproof to cars) work on an incentive basis. In other words, instead of receiving a normal hourly wage, they earn a flat rate for every car they rustproof. The more cars they rustproof, the faster they work, the more money they make. So, few cars are rustproofed with any great attention to detail. Consequently, the critical areas that are most susceptible to rust are usually poorly done. The end result: ineffective protection.

Secondly, an alarming number of companies that provide the rustproof coatings and administer the warranties have gone out of business. If your car is rustproofed by a company that no longer exists and you have a rust problem, you're out of luck. Also bear in mind that most rustproof warranties very specifically state that they do not cover surface rust. Only rust that is deemed to be from the inside out is covered. This is a very ambiguous clause. Without the help of a metallographic expert and New York City attorney, how could anyone determine that a perforation was caused by internal rust rather than surface rust?

As far as paint sealant goes, I assure you that it is little more than an expensive wax. It doesn't really do much to protect your paint. As for fabric guard, it would be just as beneficial to buy a can of Scotchguard and protect the interior fabric yourself. Please don't waste your money on any of these overpriced facades to protect your car.

[NOTE: In recent years, some marketing genius has come up with what is called "electronic rustproof." Basically, it is some kind of electronic gizmo that is installed under the hood of your car. The theory is that somehow (it's been explained to me a hundred times, with all kinds of scientific documentation, but I still don't believe it!) a low-current electrical impulse prevents rust from oxidizing on your car. The bottom line is this: Forget it.]

If you encounter a dealer who typically adds these options to his cars prior to the sale, insist that you will not, under any circumstances, purchase anything other than a factory-fresh car *without* after-market protection packages. If the salesperson or sales manager tries to persuade you to take the car with the protection package because they're going to throw them in for *free*, don't fall for it. Be firm and tell them emphatically that you will take your business elsewhere if they try to sell you a car that has those extras on it.

After-Market Accessories

As if buying a car weren't enough of a chore, the plot thickens. Like protection packages, most dealers offer dozens of very appealing accessories to dress up and personalize your new car. Everything from alloy wheels to CD players is available. Dealers offer floor mats, pinstripes, ski racks, sunroofs, center arm rests, beverage trays, graphic equalizers, and so on.

Let me clarify an important point. When I refer to after-market accessories, I am referring to dealer-installed options, not accessories that are factory installed. All options that are factory installed are clearly itemized on the factory sticker.

Just like rustproof, paint sealant, and fabric guard,

dealers enjoy significant markups on these items *and they also add accessories to cars before the cars are sold*. They do this not only to pump up the list price and increase markup, but also because they know that desirable accessories can get customers excited and start the buying juices flowing. It gives the dealer an emotional edge, but it is a detriment to the buyer. It is very difficult to effectively negotiate a deal on a new car that has dealer-installed accessories. The markup on most of these options will be substantial. Considering that different dealers may charge more or less than a competitor for the same accessory, it can create confusion and make the negotiating process more difficult.

Whether or not you choose to accessorize your car is a personal decision. However, let me give you some tips that will save you a lot of money. First of all, *never* attempt to negotiate a deal on a car that is accessorized, even if you intend to purchase the accessories anyway. If the car dealer claims that he is going to throw them in or discount them drastically, do not fall for it. Just like the protection packages, accessories are a method to increase dealer profit and to camouflage the *real* sticker price of the car.

If you choose to purchase accessories for your new car, I strongly suggest that you *do not* include them in the negotiating. If there is any way possible that you can pay for these accessories out of pocket, you would certainly benefit. Even if you have to use a Visa or Mastercard, it would be better than including them during the bargaining. When your bargaining begins on a new car, be absolutely sure that you are negotiating on the specific car and the specific factory options that you have chosen. Look at the factory sticker or the manufacturer's suggested retail price (MSRP). The car that you want to buy has to be factory fresh, without any dealer-installed options, in order to ensure the best possible negotiating edge. Remember that factory stickers are mandated by the federal government and must appear on *every* new car and light-duty truck sold in the United States.

If you choose to purchase accessories for your new car, do not let the salesperson know that you are even mildly interested in any additional options. Only after you, the salesperson, and the sales manager come to terms on the price of the new car, do you begin the process of bargaining on accessories. You should be happy to learn that not only cars are subject to the horse-trading business—accessories are too.

After you have successfully negotiated an acceptable price on the car itself, then and only then will you be prepared to deal on accessories. In Chapter Ten, Formulating the Plan, you will be instructed to ask various salespeople for a price list of available accessories as you shop around gathering information, prior to entering the negotiating arena. Most dealers have an accessory menu or the salesperson can escort you to the parts department where all available accessories and their retail prices are openly displayed. You should carefully consider the value of any aftermarket accessories. Typical markup on dealer-installed accessories is between 25 percent and 50 percent. With a little perseverance you should be able to buy almost any accessory for 5 percent to 10 percent over dealer cost. Remember that after you and the sales manager have agreed on the price of the car, the dealer has an additional opportunity to increase his profit. Even if he only makes a 5 percent profit on the accessories you purchase, it is pure gravy to him. I would strongly suggest that before you purchase any dealer-installed accessories you check out the specialty shops that sell those accessories. Many times these specialty shops sell comparable options for a fraction of the cost. But beware, the quality of accessories at a specialty shop may not be equal to a car manufacturer's.

If you decide to purchase any dealer-installed accessories, this is how you do it. After you have consummated a deal on a new car and all of the paperwork is signed, the salesman will turn you over to the finance and insurance (F & I) manager. He will review the paperwork and try to pitch you on financing through the dealer and buying an extended warranty and dealer-installed accessories.

You must predetermine what options you wish to purchase and their total retail price. Let us assume that you are interested in dealer-installed options that have a total retail price of $500. Offer the F & I manager $300 for them (approximately 40 percent of retail). He will protest and swear that you must be crazy, but when all is said and done you should be able to purchase *any* dealer-installed accessories for a minimum of 25 percent off retail. Stick to your guns and insist that you will not buy any accessories for full retail price. Insist on a fair and reasonable discount.

Rebates and Reduced Interest Rates

With little doubt, I can safely say that the most confusing and misleading advertising promotions are focused around rebates and reduced interest rates. For several years now the consumer has been bombarded with every conceivable incentive to stimulate the buying juices.

There are so many programs and complex guidelines attached to these incentives that sometimes I think that even the dealers themselves get a little confused. Some incentives are direct from the manufacturer. Some require dealer participation. Yet others are totally generated by the dealer and more often than not have little value.

There are several very important things to consider when trying to measure the value of an incentive. First and foremost, *read the fine print*. You must determine which of the four basic types of incentive programs is being offered. 1) Is the incentive program totally sponsored by the manufacturer? 2) Does the dealer have to participate in the program by helping to contribute to the incentive? 3) Is the program strictly a dealer incentive without any factory assistance? 4) Is it a direct-to-dealer cash incentive that is not offered to the consumer?

You must ask the salesperson to show you the guidelines of the particular incentive program you are evaluating. If printed material does not exist, then the program may be an in-house dealer incentive and I strongly urge you to avoid it no matter how tempting it may sound. There is no such thing as a free lunch. Don't become a victim of advertising hype. There has *never* been a dealer-to-buyer rebate or reduced interest rate program that translated to a good deal for the car buyer.

All factory rebates and reduced interest rate programs are outlined very specifically to dealers. Insist that the dealer show you the program guidelines printed by the manufacturer. An entire book could be written solely for the purpose of evaluating and defining each of these incentives. Just remember to read the program thoroughly so you can objectively weigh the merit of the incentive in relation to the overall deal. A rebate or reduced interest rate all by itself does not make a deal on a new car a good deal.

Also remember that automobile manufacturers are in business to make money. Incentives are designed for one purpose only—to sell cars. Many of the incentives offered by manufacturers may be on cars that simply are not selling. Don't buy an inferior car just because you're dazzled by a $2,000 rebate. Saving $2,000 on a car that traditionally has had a poor resale value can cost you $4,000 when it comes time to trade the car in. This does not mean that rebates and reduced interest rates are never offered on high-quality cars. But most of the time their purpose is to move slow-selling vehicles. Rebates and reduced interest rates are fringes. Although they do have obvious value, I assert that they should not play a significant role in the overall buying decision. If you do purchase a car offering a rebate, exclude it from the negotiating process. Apply the rebate to your total down payment. This will help to reduce your monthly payment or total out-of-pocket cash.

It would be most helpful if you can get your hands on a recent copy of *Automotive News,* a weekly trade publication that is packed with valuable information. The second-to-last page outlines all car manufacturer's current rebates, reduced interests rates, and dealer incentives. This publication is not normally available to the general public. Try your local library or a magazine store that carries trade publications.

Financing Through the Dealer

There are all kinds of misconceptions related to financing a car. Sometimes the financing game can be as complex as car buying itself. I can clear up 90 percent of all financing myths by reminding you that mathematics is an exact science. Finding the best financing source for a new car is as simple as addition, subtraction, multiplication, and division.

If you are skillful enough to negotiate an exceptional deal on a new car, you must also be sure that you don't lose that savings in the dealer's finance office. The only viable way to find the best financing available on a new car is to compare annual percentage rates (APRs). I don't care what the dealer tries to tell you to camouflage the real cost,

APR is everything. A $10,000 loan for a 60-month term increases by more than $250 for each percentage point the annual percentage rate increases. If a car dealer's F & I manager persuades you to finance through the dealer because his rate is only a *little* higher than a local bank, remember that a little higher can translate to a lot of money.

Most dealerships have full-time F & I managers. When you purchase a new car, the salesman is required to turn you over to the F & I manager (sometimes called the business manager). It is his job to induce you to finance through the dealer rather than go to your own bank or pay cash. These F & I managers are perhaps the strongest salesmen in the dealership. Their ability to close car buyers and convince them to finance through the dealer is most impressive. The F & I manager's special computer software can dazzle you with all kinds of interesting information.

Before venturing out to buy a car, you need to check with several local banks and your local newspaper to get a feel for current interest rates. Also check with your credit union if you belong to one, because nine times out of ten, their rates will be lower than most banks'. I'm not telling you to completely discount the idea of financing through the dealer, but remember that a dealer is a middleman between you and the bank. Dealers don't offer financing because they're good samaritans. They make a great deal of money through the financing office. If the dealer rate is competitive with local banks, then consider financing through the dealer.

A word of caution. If you are one of the less fortunate people when it comes to credit ratings, you may be forced to finance through the dealer because financing directly may not be possible. It is entirely possible that even though your local bank may refuse to approve a car loan, the dealer may be able to get an approval. This is because the volume of business that dealers provide to banks gives them a tremendous amount of clout when trying to finance a less desirable customer. Simply put, you may be forced to use the dealer's financing if you have poor credit.

Most salespeople will try to feel you out during the sale to find out if you are going to finance through the dealer. You should *always let the dealer assume* that you will be using their financing so they will also assume more profit in the total deal. This will allow you to negotiate a lower price because the dealer will presuppose that whatever he gives away on the price of the car he will make up in the finance office.

Extended Warranties

Manufacturers' warranties vary considerably from one car maker to another. Unfortunately, the real value of warranties has been clouded because one really needs to read between the lines to clearly understand the value of a warranty.

For example, all manufacturers' warranties are qualified by time (36 months) and miles (36,000 miles). Thus, a 36-month/36,000 mile warranty covers *all* defects in workmanship and repairs from the front bumper to the rear bumper, unconditionally, right? Maybe. Maybe not.

There are really three basic kinds of warranty. Bumper-to-bumper warranties, power-train warranties, and those that incorporate a little of both. A bumper-to-bumper warranty simply means that for all practical purposes, just about every nut and bolt in your car is covered for the specified period. On the other hand, a power-train warranty usually covers only the engine, the transmission, and the drive axles. Many times power-train warranties have deductibles attached to them. The only *real value* in a warranty is a bumper-to-bumper warranty.

As you do your research to determine which car you will ultimately purchase, the quality of the factory warranty is important. However, don't assume that a 7-year/70,000-mile warranty is twice as good as a 3-year/36,000-mile warranty. If the 7-year/70,000-mile warranty really is a combination of a 3-year/36,000-mile bumper-to-bumper warranty and a 4-year/34,000-mile power-train warranty, the additional value is negligible.

It is very common for car buyers to purchase extended warranties from car dealers. These extended warranties literally lengthen the factory warranty to a longer time period and a higher mile limit. Extended warranties come in many different forms and with varied coverages. Some are virtually the same as factory warranties, as far as overall coverage, and some are completely different. Without question, car dealers realize considerable profit from selling extended warranties. Dealer markup can be substantial. Extended warranties are offered by most manufacturers as well as by a dozen independent companies who

go like the rain and the wind. There
al things to consider when deciding
or not to purchase an extended war-
Bear in mind that a warranty that costs the
car dealer about $225 can often be sold for over
$1,000—a pretty fair profit considering that other
than a small percentage of claims, the only tangi-
ble cost to the dealer is administration.

First and foremost, don't even entertain the
idea of an extended warranty unless it is offered
and guaranteed by the *manufacturer*. After-market
warranty companies have had a history of insta-
bility, and the marketplace is so volatile that it is
risky for you to consider *anything* but an extension
of the factory warranty, by the factory. Should
you relocate to a different area, or experience car
trouble while out of town, it would be uncertain
whether or not a dealership other than the one
that sold you the warranty would honor it. Secondly, unless the extended warranty is clearly defined as a bumper-to-bumper warranty, don't even
consider it.

Ask yourself these questions. How many miles
a year do you drive and how long are you going
to keep your car? If you drive 15,000 miles a year
and intend to keep your car for 5 years, a typical
36-month/36,000-mile factory warranty will only
protect you for about two and a half years. You
will not be covered for repairs for 30 months and
39,000 miles of ownership. That's a long time to
be on your own.

How many electronic gizmos does your car have?
Is it basic transportation or does it have all the bells
and whistles? Cars with power sunroofs, power

windows and locks, air conditioning, cost a considerable amount of money bumper-to-bumper warranty will cove options. Find out if the warranty is This can add to the car's resale value, particularly if the car is traded sooner than anticipated.

The bottom line is this: Read the warranty carefully so that you fully understand what is and what is not covered. If, after evaluating all factors, you can buy a quality manufacturer's bumper-to-bumper warranty that will protect you for the entire period of ownership, it is probably a prudent decision to do so.

Just like after-market accessories, you can also negotiate the price on an extended warranty. Let me give you some general price guidelines. Markups can vary dramatically from dealer to dealer and manufacturer to manufacturer. A general rule would be to assume that most warranties are priced 75 percent to 100 percent over dealer cost. Now bear in mind that the dealer is entitled to make a profit on the sale of warranties. Your individual negotiating skills will determine the price that you ultimately pay.

Commitment—The Dreaded Word

When you are looking at new cars, whether you're shopping seriously or just kicking tires, all salespeople are trained to make you a *today* buyer, whether you like it or not. Sales managers hammer salespeople constantly to persuade shoppers to make a "commitment." Let me define commit-in car dealer's terms: If you are willing to buy a car *for any* price today and sign the retail buyer's order *today, you've* made a commitment. It doesn't matter if you *want to buy* a $15,000 car for $5,000, all that matters *is that for some* undisclosed price, you will commit to a *purchase* today. The logic is this: If the salesperson can get you to commit to buying a car at any price, once the commitment is out of the way, he and the sales manager will work on you to increase your commitment to an acceptable amount. Thus they will sell you a car today.

Two of the classic phrases used by salespeople are the old *"If I could, would you?"*, and, *"What do I have to do to sell you a car today?"* This is called putting the ball in the customer's court. Never allow yourself to be trapped by this delusion. If a salesperson can convince you that you just might be able to buy a car for a ridiculously low price if you sign an offer today, you are in big trouble. It is the surest way to be trapped into a very uncomfortable situation. Control is everything. If the salesperson can get you to make a commitment—any commitment—before you are prepared to enter the negotiating arena, you may become vulnerable and make a decision that you will regret.

Use all of the information in this chapter to your advantage. Remember that after you have negotiated an exceptional deal on a new car, it is equally important to maintain your victory by further negotiations on accessories, financing, and warranties. Do not be coerced into any decision that does not comply with what I have divulged in this chapter.

5

Beware of the "Selling System"

If you haven't figured it out already, let me be the first to tell you that the evolution of the car dealer has made buying a new car more complicated than ever. Brace yourself, because it gets worse!

Over the years, a number of enterprising entrepreneurs have developed various "selling systems" to help car dealers cope with the challenge of increasing volume, and most importantly, of maintaining an above-average gross profit. These selling systems are concepts that are developed, tested, fine-tuned and ultimately sold to car dealers by independent sales consultants.

Selling systems are nothing new to the sales game, nor are they unique to the automobile business. Hundreds of consultants across the country sell their systems and services to just about every industry that employs a sales force. It is no secret that any sales organization that uses a very structured, methodical selling system enjoys greater success and higher profits. Most sales managers agree that a selling system that is incorporated into a dealership's business plan will result in higher profits and greater volume.

So what exactly is a selling system? Basically, it is a technique of selling that accomplishes three important goals. Salespeople are taught to sell every prospect the exact same way. It is a step-by-step presentation that allows the salesperson and sales manager to *control, confuse, and consummate*. Let me explain.

A dealership that uses a selling system programs salespeople to follow a very strict, step-by-step presentation, called the road to a sale. Typically, the steps are: meeting and greeting, qualifying, walkaround presentation, demonstration ride, service department tour, trade-in evaluation, and closing the sale. All of these steps are orchestrated in a methodical, rehearsed fashion. This theory (which has proven to be effective at dealerships day in and day out) is based on the premise that if a complete sales presentation is sequenced in a certain way, prospects can be closed much easier. Trust me—it works.

Survey after survey has supported the fact that car buyers are much more likely to purchase a car from a salesperson who establishes a strong rapport with them. A step-by-step sales presentation accomplishes exactly that. Let's examine the control, confuse, and consummate system.

Control

I've already cautioned you in previous chapters that control is everything. If the salesperson controls you, you lose; if you control him, you win. It's that simple. A structured selling system helps a salesperson establish control because he doesn't give you an opportunity to interrupt his presentation. He defers all inquiries until he has completed the road to a sale process. He simply does not give you a chance to disrupt his game plan. He overwhelms you with professionalism and accommodation, but beware—his ultimate goal is not to be your friend, but to earn a commission. If you encounter a salesperson who tries to postpone answers to your direct questions, he is probably trying to follow a very structured selling system. Beware.

Confuse

This part of the selling system is a little more complex. Most systems use a worksheet to help keep things complicated. One of the most popular worksheets is called the four-square system. This worksheet literally has four squares on it. The squares represent selling price, trade allowance, down payment, and monthly payment. At the top of the worksheet there is space for your name, address, phone number, social security number, trade-in information, and so forth. Below this information are the four squares.

This is how the car dealer uses the four-square worksheet to confuse you. After a salesperson has given you a complete presentation, it is time to sit down at his desk and put some figures on paper. The salesperson will ask you a series of *qualifying questions* to ascertain if you are ready to make the buying decision. A typical question is, "Other than price, is there any reason why we cannot sell you a car today?" The purpose of this question is obvious. The salesperson is *really* asking you, "If my manager can put together acceptable terms (price, payments, trade-in allowance, and term of loan) will you own the car *today?*" Any answer, other than yes, will cause the salesperson to backtrack and try to find out what other objection he needs to address so that he can sell you a car today.

As soon as the salesperson convinces you to buy today, on "your terms" (or so you think) he will huddle with the sales manager to get "starting figures." When he returns to his desk, he will show you the figures on the four square worksheet and his pitch will be something like this. "I spoke to my manager and his proposal is really terrific! The selling price of the car you're interested in is $14,350; your trade-in is valued between $4,500 and $5,000; you will need to come up with a down payment of around $2,500; and your monthly payment will be between $325 and $340." Whew! Did you catch all that? He will then sit back, stare you right in the eyes, and not say another word until you react.

Most people react to these starting figures with a series of questions such as, "How much are you discounting the new car? Why are you ranging the value of my trade-in? Why do I have to come up with a down payment? Why are you ranging the monthly payment? How long is the finance term? What is the interest rate?" Can you see how the four-square worksheet can confuse you?

There are so many unanswered questions that each significant point becomes diluted.

The salesperson's ultimate goal is to persuade you to focus on payments and payments only. The car dealer maintains a strong position of control if they can keep you directed toward payments. If he can accomplish this and convince you that the monthly payment meets your budget and that it is *really* the only measure of a fair deal, it's all over but the shouting. If you agree to the terms and agree to buy the car, he will then try to convince you to take delivery immediately, thereby consummating the deal.

Consummate

When you purchase a car and take delivery the same day (sometimes the same hour!) this is called a spot delivery. Remember that a car sale isn't a sale until you physically take possession of the car. Now why do you suppose a car dealer would want you to take delivery so quickly? Because quick, impulsive buying decisions tend to cause Buyer's Remorse. And Buyer's Remorse translates to unconsummated deals. By attempting to deliver the vehicle as soon as possible, ownership takes place long before Buyer's Remorse can set in. Thus, consummating the deal ensures that the cash register will ring long before the ether wears off!

It is important for me to clarify something. I don't mean to get philosophical (well, maybe I do), but there is a very significant point that needs to be addressed. A car dealer, just like every other business owner in the world, has a legal, moral, and ethical right to make as much profit as is humanly possible. There *is* a fine line between deceptive and creative selling. A car dealer is entitled to ask you to pay full list price, or over list price, for any car on his lot. On the other hand, you have the right to try to purchase that vehicle for as little as is practically possible. Alas, the wonderment of capitalism!

As I've stated earlier, car buying is the purest form of horse trading in the world. There is nothing ethically wrong with any businessperson incorporating a selling system into a sales presentation to maximize profitability. To capsule a phrase, "Let the buyer beware." The only way for you to ensure that you will not pay too much for a new car is for you to be one step ahead of the car dealer. What truly is the measure of a fair deal? Perception is reality. It may be price, it may be the way you were treated, or it could be your overall buying experience.

I know an awful lot of first-class, highly ethical car dealers who use a selling system to increase profits. These dealers treat their customers like gold. They truly care about their customers, but they also care about the bottom line. That they wish to increase profits by outwitting the car buyer at the negotiating table doesn't make them less than scrupulous. In fact, I can assure you, with little exception, that most car dealers today are extremely sensitive to the needs of their customers. Again, they have a right to make as much profit as is legally possible.

Let me give you a couple of tips to ensure that you don't get caught in the control, confuse, and consummate trap. First of all—and I will repeat this time and time again throughout this book— you cannot even begin to enter a car dealer's showroom until you are 100 percent prepared. You must do your homework and research long before you enter the negotiating arena. When you do come face to face with a salesperson, there is nothing wrong with him giving you a first-class product presentation. But stay alert and don't fall prey to making an emotional decision rather than a logical one. Be acutely aware of worksheets, whether they are a four-square system or not. Worksheet magic can be a car buyer's worst enemy.

Secondly, when you do ultimately negotiate a deal on a new car (I'll get more specific about negotiating later on), do not, under *any* circumstances, take delivery of your new car the same day you buy it. Wait at least forty-eight hours to be sure that you made the right choice at the right price, and that your purchase was a logical one. If the car dealer tries to convince you to take delivery immediately—no matter what his reasoning—don't do it.

Thirdly, during the shopping process (which will be outlined in great detail in Chapter Ten, Formulating the Plan) do not allow a salesperson, sales manager, or anyone else to pressure you into making a buying decision until you are prepared to do so. If you feel you are being pressured, leave the showroom immediately and exclude this dealer as a viable contender for your business.

6

To Trade or Not to Trade

Perhaps the single most perplexing, frustrating, and misunderstood piece to the car-buying puzzle is the trade-in. Most people who buy new cars not only want to achieve the lowest possible price, but they also want to be assured that they obtained the highest trade-in allowance.

Understanding the trade-in is a complex and critical issue. It is essential for you to determine its *real* value before you can even begin to enter the negotiating arena. You must also discard the preconceived ideas that have anguished you in the past. Let us focus on some key points that will allow you to establish a fair value for your trade-in and then realize that fair value at the negotiating table.

A used car is a unique and individual commodity. No two are exactly alike. Each car commands a different value and to arbitrarily generalize that all 1989 Chevy Caprices are worth about the same money is a dangerous misconception. A flawless, foolproof method of establishing a value for a used car simply does not exist. Remember that when you trade your car in to a dealer, you are in effect selling the car to him. Therefore, the *real* value is somewhat subjective. You've heard the adage "One man's treasure is another man's junk." The value of your trade-in is negotiable, just like the price you pay for a new car.

The good news is that if he is properly handled, a car dealer can be persuaded to give you a lot more for your trade-in than he would like to—if you do your homework and carefully plan your strategy. Before we establish some sound, sensible methods to help you find out what your car is really worth, let us remove some common misconceptions.

Please do not look in the classified want-ads of your local newspaper trying to find cars similar to yours. *No two used cars are alike.* You cannot look at retail asking prices of *similar* cars to establish a value for yours. Secondly, many car buyers refer to "blue book" or "black book" prices. These books are a viable source, if they are used as a *guide* and if you fully understand how to use them. But they can be very misleading in the hands of a novice. Let me appeal to your logic for a moment. It is my strong belief that blue books and black books are deceiving because the only true value of any value on this earth is what someone is willing to pay for it. Books do not buy cars and they do not write checks. If the editor of a particular book is not willing to pay you what he says your car is worth, then what is the validity of his claim?

There are many things that have a bearing on the value of a used car. Geography is one factor, believe it or not. You can get a lot more for a convertible in California than you can in Minnesota. And you can get considerably more for a four-wheel-drive truck in North Dakota than you can in Florida. Time of year, miles, color, resale value, market conditions, and options all play a key role in establishing a true and fair value on a used vehicle.

Supply and demand is one of the most important components in establishing a real value for a used car. In some instances your car can be worth more in six months than it's worth today. A car's overall resale reputation in the marketplace is a key factor. Some new cars just do not sell very well as used cars. Sometimes a certain stigma exists that negatively affects the value of a particular model car. The list goes on and on.

At this point I would like to define two distinctly different automobile terms that relate to the value of a used car—actual cash value (ACV) and gross trade allowance (GTA). It is important for you to understand the difference in these two terms. The ACV and GTA both refer to the value of your car, but they are totally different figures. To confuse the two, or to be misled as to which is which, can be costly.

Let me define the difference between ACV and GTA by formulating an example. Assume that you have decided to trade in your car and purchase a new car that has a factory sticker price of $15,000. After diligent negotiating you are able to work a deal for $10,000 cash difference, plus tax, license fees, and title. More specifically, the value of your trade-in is $5,000. Right? Wrong, *expensively* wrong. Every new car has a dealer markup built into the sticker price. This markup is different from one model to another. Some markups are less than 10 percent; others more than 20 percent. The illusion here is that by the dealer allowing you a gross trade allowance of $5,000, the average car buyer assumes that the GTA is the real value of the trade-in.

What really happened in the above transaction is that you received about $3,500 actual cash value for your trade-in and a $1,500 discount, which you would have received anyway if you had negotiated a deal on the same new car *without* a trade-in.

Let me restate this one more time because it is important. If you buy a car with a sticker price of $15,000, and your trade-in allowance is $5,000, the GTA (which is a *combination* of ACV and discount) is $5,000, but the real value of your car, the ACV, is $3,500.

Understanding the mathematics of ACV and GTA is an extremely important point. Every deal, on every new car, at every dealer from New York to Los Angeles, distinctly separates the actual cash value and the discount on each new-car transaction. Salespeople don't tell you this, but it is a reality.

The real value of your car, the ACV, is $3,500. The gross trade allowance is $5,000. Never lose sight of what this means. To understand that the GTA is a combination of the ACV and the discount is a critical fact in the overall strategy used in negotiating. It is a basic fundamental that will dramatically affect your ability to intelligently negotiate a better deal when you're sitting across from a salesperson at the bargaining table.

Car dealers would like to own all trade-ins for what they consider to be wholesale value. Actual cash value and wholesale value mean basically the same thing because they are the same number, just different terminology. You should know that the profit that can be realized on used cars is even greater than it is on new cars. To more clearly understand how a dealer arrives at an ACV, examine the following numbers.

If a car dealer feels that your trade-in can be sold to another customer for $6,000, this is his logic. Do not take these figures at face value; they are used in concept only. The dealer takes the $6,000 retail figure and subtracts about $2,000 markup from it. He also deducts whatever he feels the total reconditioning will cost. For example, can the dealer simply wash and wax your car and put it in the front line or does it need tires; are there any dents that need to be fixed, or is there a cracked windshield; does it need brakes or a muffler? Let's assume that he estimates that it will cost about $350 to make your car "front-line ready." This is how the math works.

Your trade-in has an intended retail value of $6,000. He subtracts the $2,000 markup and the $350 reconditioning, and another $150 "just in case." That makes the actual cash value, or wholesale value, $3,500. Is that a fair figure to you? Maybe it is, but in many cases it is not. If you are negotiating a deal on a car with considerable markup, the dealer could show you on paper a gross trade allowance of almost $6,000, which is a retail price for your used car, and create the illusion that you were getting a fair deal. But in fact he would be appraising your trade for wholesale or less.

The lesson to be learned is simple. One cannot negotiate a fair deal on a new car without understanding the ACV of the trade-in and the dealer markup on the car. The key to negotiating an exceptional deal on a new car is to achieve a gross trade allowance that consists of the maximum actual cash value and the maximum discount. Another thing to keep in mind is that many trade-ins are not worthy of reconditioning and resale. These "wholesale pieces" are usually brought to auctions where they are purchased by junk dealers who strictly deal in cheap cars. It is important for you to objectively evaluate whether or not your car is one that will be retailed.

Many car buyers think that they have the system figured out. By selling the trade-in outright and purchasing the new car on a "clean-deal" basis, they are virtually getting the best of both worlds—the maximum amount for the trade and the highest possible discount. In theory, this logic makes a great deal of sense, but it may not always work in practical terms. Consider the following.

If you sell your trade outright, will it create a transportation problem for you or your family? Will it force you to buy a new car more impulsively because you cannot be without a car? Will you pay more sales tax on a clean-deal transaction because most states calculate sales tax on the cash difference? Do you still owe more on your trade-in than it is worth? If so, you will not be able to roll over the excess amount owed into a new car loan if you don't trade the car in to the dealer. How much is your time worth? Do you want to incur the expense of advertising your car in the classifieds, accepting phone inquiries, and setting up appointments for people to inspect and drive your car? Is your trade desirable enough to ensure that you will realize as much for it by selling it on your own as you would by trading it to the dealer? If there is an outstanding balance on your trade-in, the bank will not release the title until the balance is paid in full with certified funds or until your personal check clears (which takes about ten banking days).

Starting to get the picture? Selling your trade-in on your own can be a painstaking and disappointing venture. I will show you how to trade your car in to the dealer, to circumvent all of the problems encountered when attempting to sell a car outright, and to still maximize your overall trade allowance. Your trade-in can be a very important tool during the negotiating process. You would be amazed at how many times a car dealer will stretch the appraisal on a used car just to make a deal on a new car. I guess that you could call it immediate gratification (the sale of a new car) for future aggravation (owning a trade-in for too much money).

Here are a few pointers on how to get the most for your trade-in. There are only three kinds of used cars in the world: cream puffs, plain janes, and roaches. (Please excuse the automotive slang.)

The way you maintain your car has a dramatic affect on its trade-in value, perhaps more than you realize. Most people are seriously negligent when it comes to maintaining their cars. Changing the oil every 3,000 to 4,000 miles isn't enough. You must bear in mind that when a used-car manager appraises your car, he tries to evaluate how this car was treated. He draws a lot of conclusions from appearance.

Although it is very difficult and costly to transform a roach into a cream puff, the vast majority of cars, the plain janes, can be given the appearance of being a cream puff even if they are not. Before you enter the negotiating arena it would be well worth the effort to spend a little time and money making your car "smile." You should clean your car inside and out and repair minor mechanical problems. It should be cleaned just as if you were entering it in an auto show.

You must understand that when a used-car appraiser evaluates your car, he will deduct three times the actual cost of repairs and reconditioning from the ACV. If your car gives the impression that it has been well maintained, you will realize a considerably higher trade-in value.

Many car buyers fall prey to the delusion that it is senseless to invest time and money in a car that is soon to be traded. I cannot impress upon you enough the economics of this costly misconception. A clean, well-running car will ensure that your ACV is as high as possible. If the used-car appraiser believes that your car needs a minimum amount of reconditioning and that it is a desirable used car for retail sale, his appraisal will be very aggressive.

Here are some sound suggestions on how to transform your plain-jane car into a cream puff. First of all, be sure that the engine is running properly. If the engine is skipping, or if it starts hard when it's cold, get a tune-up. If your car's air conditioning is not working properly, more often than not the system may just need a charge, which will cost only about twenty dollars. If the used-car appraiser assumes that the air conditioner is inoperable because the compressor is defective, he will deduct about $400 from the ACV.

If your exhaust system is loud, get it repaired as inexpensively as possible. Nothing annoys an appraiser as much as a car that sounds like it should be in the Indy 500. If your windshield is cracked, many times it can be repaired rather than replaced, if the crack is what they call a "bull's-eye." Again, a repair will cost about $40; an appraiser will deduct about $250 from the ACV.

Objectively evaluate the running condition of your car and repair whatever can be repaired for a *reasonable* cost. Remember what I said earlier: If your car is a roach, it would be senseless to invest a great deal of money on repairs before trading. Once you have had all of the mechanical problems taken care of, it is then time to spit-shine your trade-in.

I do not think that anything impresses a used-car appraiser as much as a clean car. Now, I am not talking about a car that has just been washed and vacuumed. When I say clean, I mean *clean*. Your trade-in should be washed and waxed. The windows should be cleaned inside and out. The dash, the rearview mirror, and the instrument panel should be spotless. The glove box should be clean and organized and the trunk should be empty and vacuumed. Clean the ashtray; shampoo the carpet to give the inside of the car a clean, fresh smell; and remove road tar and bugs with a solvent.

When you are completely done, check the car over one more time. Look at the car just as if you were a buyer and you were scrutinizing every nook and cranny. I assure you that the ACV that you realize will be the maximum for your particular car if your trade-in really smiles. If you don't feel qualified to spit-shine your car, most cities have a number of independent doll-up shops that can make your car look like new for a reasonable charge.

Now that your car is running like a top and is as clean as a whistle, it is now time to establish a fair trade-in value. I believe that there is only one viable method to determine the ACV of any used car. This method involves a little running around and a lot of patience. (How much time are you willing to spend to achieve a trade-in value that is $500 to $1,000 more than you would normally get?) Because every used car is unique, each geographic location is distinct, and dealer criteria in valuing a used car is individual, classified ads, blue books, and black books are simply not the answer. What you must do is some independent research.

First of all, every city, town, or village in the country has any number of independent used-car lots that are not affiliated with a new-car franchise. These used-car lots specialize in used cars and only used cars. Some of these used-car dealers have a diversified inventory, but you will find that usually each of them has a certain niche.

Some sell nothing but "cheapies," cars that cost under $1,500. Some sell only high-line cars or luxury cars. Others might specialize in imports. Whatever the case, you must objectively categorize your trade-in. Then you must "shop" two or three of these dealers and attempt to sell your car to them. Of course you are not actually going to *sell* your car; you are only trying to establish a real ACV. This is what you do.

Do not call them on the telephone. You must actually drive to these used-car lots so the owner can see your car. Your story is simple. Tell them that you are interested in selling your car outright. You are not going to buy another car to replace it; you just decided that you own one car too many. Make sure that you do not give the impression that you are in a financial bind. If you do, it will give the dealer the green light to try to underprice your car. Also, make it clear that you are not going to buy a car to replace it.

Remember that a used-car dealer, like a new-car dealer, would like to own your car for the lowest possible price. Ask him how much he will pay you for your car. Don't expect him to immediately give you his best offer. Whatever he says, rest assured that his offer will be 15 percent to 20 percent lower than the maximum he will pay. Negotiate the best price you possibly can. Repeat this at one or two other used-car dealers. You will quickly get a feel for the real ACV of your car.

A word of caution. Do not attempt to get the used-car dealer to appraise your car without actually making a specific cash offer. If he appraises the car, his opinion means about as much as a blue book or a black book. Appraisals are a dime a dozen, but only cash offers have real value. A fair value for your car can only be established with a *real* cash offer. Here is the curveball. As we discussed earlier, some cars have excellent resale values, while others do not. If you have a difficult time getting a reasonable offer on your car—or any offer at all—you must accept the fact that your particular car is not one of the high-demand cars and you will have to establish a value elsewhere.

Make sure that you get a commitment from the used-car dealer. After he gives you his final offer, tell him that you are a little disappointed and that you thought that your car was worth a little more. Tell him that you would like to sleep on it and ask him if he will stick to his offer if you come back tomorrow. If he will, you probably have a serious offer. Do this again with at least two more used-car dealers.

Now, do the exact same thing with two or three new-car dealers that sell the same car that you are trading. When you enter the dealer's showroom, ask for the used-car manager. He is the only one that you want to talk to. And remember, your story is that you only want to sell your car, not buy a new one. Again, if you have a difficult time getting a cash offer on your car, let this be a lesson on the importance of buying a new car that has a strong resale value.

Take the highest offer you receive and add 10 percent to it and you will have a fairly accurate ACV for your used car. When you ultimately enter the negotiating arena, you will have an invaluable advantage. Unlike most uninformed, naive car buyers, you will have a real value established for your trade-in.

If you are unable to get an offer on your car you will have no alternative but to consult a blue book or a black book as a guide. If you are unfortunate enough to own a car that simply is a low-demand car, you have no other choice. I would strongly suggest that if you are forced to use a book, look at the most consistently accurate book: the NADA (National Automobile Dealers Association) used-car guide. You can go to your local library and use their NADA book or you can purchase one at most bookstores. *Important!!* Whatever book you use, be certain that you look up the wholesale value of your car, not the retail value. Most books disclose three values: retail value, trade-in value and loan value. Loan value is the wholesale value and that is the number you are looking for. In using the NADA book, be sure to deduct for excessive mileage and to add or deduct for noted options.

7

Narrowing the Field of Choices

Fifty years ago, when one was faced with buying a new car, there were so few choices and options available that value for the dollar really wasn't the issue. In today's market, narrowing the field of possible choices can be extremely confusing. First of all, not only do you have hundreds of different models to choose from, but each year our technology makes available to us a whole spectrum of options, accessories, and fancy frills to get our juices flowing. Some of these options have practical, sensible, and safety-related applications; others are simply available to us so we can pamper ourselves.

Back in the forties and fifties, imports were virtually nonexistent. Safety didn't seem to be of particular concern and air pollution wasn't even a major issue. Comfort and convenience weren't a part of promotion or advertising. Fuel economy really didn't catch anybody's attention—as long as an automobile got you from point A to point B, that was all that mattered. Today it is dramatically different.

The automobile manufacturer of the nineties has bombarded us with every conceivable option and accessory imaginable. Advertising has become so sophisticated and convincing that we all become victims of buying beyond our means, or, worse yet, buying a product because strategic advertising has sold us a bill of goods. Buying a new dress or a color television that taxes our budget is one thing, but buying a new car that is impractical or frivolous is quite another. Before we take a close look at options and features available on the cars of the nineties, let me give you some things to consider when comparing domestics and imports.

What *is* the better choice—domestic or import? I'm probably going to disappoint you, but you're not going to get a decisive answer from me. The battle between domestic and import car manufacturers is perhaps the rivalry of all rivalries. Millions of dollars are spent on advertising each year to try to persuade the consumer that a particular car is the best vehicle for the money. How does one decide?

First of all, there is a great deal of subjectivity related to this question. I guarantee you that if you read enough buyer's guides and car books, you will encounter contradiction after contradiction. Remember that no matter how credible the source of information, a test or survey result always involves someone's opinion. I strongly believe that the perception of the individual buyer is the ultimate barometer that determines whether or not one chooses a domestic or an import. I know that this is not what you want to hear, but it is a reality and perception *is* reality.

If you have owned two or three Chevrolets and have enjoyed dependable service from them, it would be very difficult for any documentation, report, or survey to convince you to buy a Toyota. On the other hand, one bad experience with a particular car will probably blackball it from consideration for the rest of your life.

Hondas are without question one of the finest cars built in the world. Their track record for dependability, resale value, and performance is well documented and hard to challenge. Review after review in every major car magazine has applauded the Honda products year after year. Why doesn't *everybody* drive a Honda? How do other manufacturers compete with a product that has so much credibility? How could any sane, well-informed car buyer purchase any other make car? Get the point?

Try to be as objective and open minded as possible. Chapter Nine will assist you in determining your personal budget. It will also discuss functions and features, frequency of repairs and resale value. All of these factors are critical components in determining which automobile is best suited for your needs.

Let us take a look at a number of options and features available on the automobiles of the nineties and evaluate their benefits, practicalities and overall value.

Air Conditioning: Air conditioning holds its value more than any other option, year after year. If you look in a NADA used-car guide and reference the resale value of any vehicle from an Audi to a Volvo, air conditioning maintains almost 50 percent of the original value for five or six years. If you like a car with air conditioning, it goes without saying that it is a smart investment. If you don't particularly like air, remember that if the retail price is $1,000, you'll get back about $500 of your investment at trade-in time. It is an option to seriously consider.

Automatic Transmission: Like air conditioning, an automatic transmission adds a significant amount of resale value to most cars. Although it retains less than 50 percent of value, it is still substantial. I personally like to drive a five-speed because I truly enjoy *driving* a car. If you encounter a lot of stop-and-go traffic, or if your highway driving is minimal, an automatic is a sensible choice. If you live in an area that has a great deal of snowfall, a manual transmission does give you more control in inclement weather. Another factor that you may want to consider is fuel economy. Typically, the EPA rating with a manual transmission is a little higher than with an automatic. If you are undecided, fuel economy may be the deciding factor. Remember that an automatic transmission costs more than a manual to begin with and the fuel savings of a standard transmission makes the automatic even more costly.

Anti-Lock Brake System (ABS): This option certainly has become a buzzword in the nineties. ABS brakes are a safety feature to thoughtfully consider. Although it is not an option available on most inexpensive cars, I believe that it will be available on all cars in the near future. An ABS brake system will not allow the brakes to lock up even if you slam on them. With a conventional brake system, if you were to jam the brakes, they would lock and the tires would skid rather than roll. This will dramatically affect your ability to steer and control the direction of the car. With ABS brakes, even if you put both feet into the brake pedal, the tires will continue to roll rather than skid to ensure a safer, more controlled stop. ABS brakes are standard on some cars and optional on others. As an option, they can cost as much as $1,000 or more. One only needs to employ the benefit of this brake system once to make it priceless. It's a serious option to consider.

Intermittent Windshield Wipers: I think that most people are familiar with this option. A variable-speed wiper system is far superior to a conventional two-speed system. It is most annoying on a misty or drizzly day to have your wiper blades squeaking as they move across a dry windshield. This feature is standard on all but the most inexpensive cars. If it is an option on a car you are considering, I give it a high recommendation.

Power Windows, Door Locks, and Seats: Although for the most part these options are simply creature comforts, there is a practical side to each of them. Power windows give you the ability to operate all the windows in the car from the driver's seat. Many power window systems have an on/off control button that will prevent the windows from being operated. If you have children, particularly in the backseat, it would be a nice safety feature to be able to prevent them from opening the rear windows. Power door locks also give you the comfort and safety of locking all doors with the flip of a toggle—a nice option to have if you should be in an uncomfortable situation and need to lock your doors immediately. Power seats can be most comfortable for people who spend a great deal of time in their cars. They can tilt forward, back, up, and down. Changing the position of the seat with this power feature can make driving long distances much more comfortable. These features do add to the resale value somewhat, but the decision to include them on your new car should be based more on personal preference.

Child Safety Rear Doors: If you have small children who sit in the backseat and do not use a carseat, this option is an excellent safety feature. It is not complicated or costly. It allows you to flip a small control lever inside the door jam to prevent the rear doors from being opened from the inside. If you cart the kids around regularly, I would put this option on the must-have list.

Cruise Control: If you do a lot of highway driving, cruise control is a very practical and economical option. Not only does it allow you to relax a little on a long trip by allowing you to change the position of your right foot, but it is a sure-fire way to improve your fuel economy. Most of us fluctuate our highway speeds by as much as ten or fifteen miles per hour without even knowing it. This constant change in speed reduces fuel efficiency. Cruise control maintains a constant speed, thereby maximizing fuel economy. It is a must-have option for anyone who drives a lot of highway miles.

Anti-theft Alarm System: If you live in a big city where crime and vandalism are a reality, then by all means you should consider an anti-theft alarm system, particularly if you have a sophisticated stereo system or a cellular phone in your car. If it is not offered by the automobile manufacturer on the car that you ultimately choose, there are a number of acceptable after-market sources that offer excellent systems. Do a little research and ask a lot of questions before you select a system. Many insurance companies will give you a slight discount on your premium if your car has an anti-theft alarm system.

Driver's Side Air Bag: There has been a tremendous amount of controversy over the issue of air bags. I for one am a strong proponent of air bags. A couple of years ago I was involved in a head-on collision and I was not wearing a seatbelt. Although I was fortunate because the injuries I incurred were not life threatening, it took me a long time to recover. I still experience residual pain in my jaw, but am more than thankful for my good fortune. One cannot possibly fathom the magnitude of the impact experienced in a head-on collision. I believe that air bags should be made mandatory on all cars. I know that there are a lot of economic implications, but I believe that air bags used in conjunction with seatbelts and shoulder restraints will dramatically reduce the number of serious injuries resulting from head-on collisions. If it is at all possible, consider a car that has an air bag either standard or optional.

I have tried to cover options that I feel should be considered before purchasing any new car. Obviously, there are hundreds of options I did not illustrate, but most of the important ones have been examined.

It is worthy to make a couple of points about safety and crash tests. Nobody likes to be morbid, but the safety issue should be a top priority when considering a new car. The reality of crash tests cannot be ignored. Crash tests reveal statistics that can be very sobering. Before you make the ultimate buying decision, it would be worthwhile to do a little research to determine how your choices rank in safety. A publication well worth purchasing is *The Car Book*, by Jack Gillis. It is packed solid with useful information on safety, crash tests, and many other eye-opening topics. I highly recommend this book as an essential tool in determining which car is best for you.

8

One-Price Selling

There is an awful lot of hoopla today about a concept called *one-price selling*. Although one-price selling is nothing new—Volkswagen used this principle with great success in the sixties—it certainly seems to have captivated car dealers, the media, and consumers alike. *Automotive News*, the car industry's foremost trade publication, has printed article after article about car dealers who have adopted a one-price selling philosophy. NADA (The National Automobile Dealers Association) is conducting seminars to teach automobile dealers how to adopt a one-price system in their dealerships. J. D. Powers, one of the largest and most well known automotive research companies in the world, is continuously conducting extensive studies concerning the viability and evolution of one-price selling. And Saturn Corporation, a division of General Motors, has employed one-price selling since the introduction of its cars. So what's all the fuss about?

One-price selling is a system that in *theory* removes the negotiating—the part of car buying that consumers dislike the most—from the car-buying process. All cars are clearly marked with "no-dicker stickers," which presumably represent the dealer's absolute best price on each vehicle within his inventory. It sounds pretty simple, doesn't it? It may possibly be too good to be true.

For those car buyers who wish to remain uninformed and are unwilling to exert a little effort and do some research to ensure the best value for the dollar, I suppose that it is the lesser of two evils for them. However, consider the following.

First of all, refer back to Chapter Two, Timing: An Important Piece of the Puzzle. Remember the average gross profit theory? Uninformed buyers pay significantly more for new cars than informed buyers do. Therefore, if you choose to purchase a car from a one-price dealer, you're going to pay what everyone else pays for a particular model car. Translated into dollars and cents, you're going to pay too much. Removing the negotiating process from car buying may be good for the dealer and the naive car buyer, but it doesn't work for the adept buyer. Remember, your goal is to purchase the most car for the least amount of money, not to pay what everyone else pays. The only way to ensure the lowest possible price is through strategic negotiating.

Secondly, most new-car deals involve a trade-in. Do the one-price dealers also give you a non-negotiable trade-in value? You can be certain that they do. Again, trade-in values are a significant component of the negotiating process. Every used car in the world is unique unto itself. And trade-in values can vary dramatically from one dealer to the next. The only way to maximize the value of your trade-in is to include it as part of the overall negotiating strategy, particularly if your car is one that the dealer can resell as a retail used car.

Thirdly, because the rebirth of one-price selling is still a novelty, only a small percentage of car dealers across the country are employing it. Most dealers are in a wait-and-see mode. What is going to happen when a greater number of dealers (assuming that one-price selling continues to flourish) embraces one-price selling? Suppose a given geographic area has ten Chevrolet dealers and seven of them introduce one-price selling in their dealerships. If Dealer A is selling a Cavalier with specific equipment for $10,500, more than likely Dealer B will undercut his price and offer the same car for $10,450, and so on. Doesn't that get car buyers right back to the negotiating process again? Doesn't it force car dealers to abandon the one-price selling concept? Won't it motivate you to shop around for the best price?

I believe that one-price selling is a temporary Band-Aid that creates the dangerous illusion that negotiating is no longer required to ensure the lowest possible price on a new car. It is an overreaction to the never-ending cries from consumers that they hate the process by which they purchase new cars. Again, when you are finished with this book, unlike the typical car buyer, you will embrace the negotiating process with an earnest exhilaration. The most feared and misunderstood segment of the buying process will soon become your most cherished ally.

I strongly suggest that you stay away from one-price dealers if your mission is to pay as little as possible for the new car of your choice.

9

Putting It All Together

To this point, this book has been devoted to disclosing the most common pitfalls of buying a new car, to clarifying misconceptions, and to removing certain delusions. You now have at your disposal an arsenal of priceless information. But this information has little value until you come face to face with a salesperson at the negotiating table and use *everything* you've learned. But you are not quite ready yet. All of the planning, number crunching, and research we discussed in earlier chapters will now become the focal point.

Before you enter the negotiating arena you must take some time to assemble your plan. You must organize your information and make some decisions. You must take out your calculator and crunch numbers. Remember, when you enter the dealer's showroom to negotiate a deal, you cannot leave *anything* to chance; plan your work and work your plan.

Let us now outline, step by step, the homework you need to prepare before you enter the negotiating arena. In this chapter we will discuss select-

st car for the least money. We will
your personal budget. And we will try
your possible selections down to three
odel cars. You will need to invest about
five dollars to purchase a car buyers' guide that
reveals dealer invoice on virtually every car and
every model available for sale in the United States.
There are several different books available; be
sure that the one you choose is a current edition.
I highly recommend the *Edmund's* series of books.
They publish three different invoice price books—
one for vans, pickups, and sport utility vehicles;
one for import cars; and one for domestic cars.
Also, *Consumer Reports* prints several books that
give accurate dealer invoice information.

Selecting the Most Car for the Least Money

It would be very easy for me to attempt to pressure you into considering a particular manufacturer or model based on my opinion. Although
I have spent more than seventeen years in the
automobile business and feel qualified and justified in doing so, I will let you make the decision
based on *your* common sense, not mine. I strongly
suggest that if in the past you have been a flag-
waving, domestic diehard, now is the time to forget
about World War II. If your research and budget
suggest that you should buy an import, then do
so. On the other hand, the import purists should
recognize that domestic manufacturers have made
great strides to improve their cars. Domestic cars
have come a long way in the last decade and

many can match the credentials of the finest imports.

There are four basic categories to review when considering the most car for the least money. They are: personal budget, functions and features, frequency of repairs, and resale value.

Personal Budget

This topic speaks for itself. No matter how much research you do, no matter how good a negotiator you are, and even if you have the most foolproof buying system in the universe, everyone has a budget. Few of us have the luxury of buying whatever we want, regardless of price. With the budget consideration in mind, the first thing you must do is find out just how much you can afford to spend on a new car.

Most people equate their maximum budget to a monthly payment. Let's face it, the average consumer may not be sure if they can afford to spend $15,000 on a new car, but can quickly tell you if they can afford $250 a month. With this monthly budget in mind, let's consider the six things you need to know before you can establish a budget. They are:

1. Maximum monthly payment
2. Term of the loan (three, four, or five years)
3. Annual percentage rate
4. Local sales tax rate
5. Down payment
6. Closeout balance on trade-in (if applicable)

(Note: If your trade-in is not owned free and clear, the balance owed to the lending institution is called the closeout balance.)

Example: Let's assume that you carefully evaluate your financial situation and feel that you can afford to *comfortably* spend $250 per month on a car payment. We will also assume that you are willing to finance the car for four years and that your local bank can give you an interest rate of 12.5 percent. Your sales tax rate is 7 percent, you're willing to put $1,200 down, and the closeout balance on your trade-in is $750.

Look at the Appendix on page 158 and locate the Monthly Payment Factor Chart. Cross-reference 12.5 percent with 48 months. You get a factor of .02658. Now pay very close attention. Divide your ideal monthly payment, $250, by the factor .02658. That equals $9,405.57. Add your $1,200 down payment to this figure: $9,405.57 plus $1,200 equals $10,605.57. Subtract your closeout from this figure: $10,605.57 minus $750 equals $9,855.57. Now remove the sales tax from this figure by dividing $9,855.57 by 1.07 (if your tax rate is 8 percent, divide by 1.08; if it is 6 percent, divide by 1.06, and so on): $9,855.57 divided by 1.07 equals $9,210.81. Bingo! This is a *very important* number: $9,210.81 is the *maximum cash difference* you can pay for a new car. What does all this mean? If you are looking at a car with a sticker price of $12,500, in order for you to achieve the desired payment of $250 per month, all variables considered, the dealer will have to allow you a GTA (Gross Trade Allowance) of $3,289.19. (Remember that the GTA includes the ACV of your trade-in and whatever discount you negotiate.)

Knowing what your maximum budget is *before* walking into a car dealer's showroom is a key component of the overall negotiating process. This information will not only save you a lot of time, but it will prevent you from chasing deals that are mathematically impossible to achieve. Smooth-talking salespeople will not be able to persuade you to buy more of a car than you can afford if you stick to your guns.

If you are fortunate enough to be able to pay cash for a new car without any consideration for financing, I congratulate you. You are part of a very elite group. All you will need to do to determine your budget is to take the maximum amount of cash you can spend, subtract the closeout balance on your trade-in (if applicable), and remove the sales tax to establish the cash difference. For example, if you can afford to spend $10,000 cash on a new car, you owe $500 on your trade-in, and your sales tax rate is 6 percent, this is the calculation: Subtract the closeout of $500 from the $10,000, which equals $9,500. Remove the sales tax by dividing $9,500 by 1.06, which equals $8,962.26. This tells you that no matter what price car you consider, your cash difference cannot exceed $8,962.26.

Now that we have played with some hypothetical numbers, let us determine your real budget. Refer to the Budget and Payment Worksheet in the Appendix and fill in the following information:

1. Maximum monthly payment. As we discussed

earlier, only careful thought can accurately determine what monthly payment will be acceptable to you. Carefully consider a payment that will *comfortably* fit your budget. Enter the payment on the worksheet.

2. Desired term in months. Decide what term of payments makes the most sense for you. Remember that the number of miles you drive per year should be considered. Also, the longer the finance term, the higher the price of the new car you can purchase. Please bear in mind that the longer the finance term, the more overall interest you will pay. Enter the term on your worksheet.

3. Projected interest rate. You need to make several telephone calls to determine the prevailing interest rates. Call at least three banks and inquire about auto loan rates. Don't forget that if you are a member of a credit union, you should inquire about their rates, since they are most competitive. Also, remember that when it comes to interest rates, the only real measure is APR (annual percentage rate). Enter the interest rate on your worksheet.

4. Monthly payment factor. Go to the Appendix and find the Monthly Payment Factor Chart. Cross-reference the APR and the term in months. Enter the factor on your worksheet.

5. Sales tax rate. Enter the sales tax rate for your area on your worksheet.

6. Maximum down payment. Carefully evaluate your personal financial situation and decide if you are able to put a down payment on your new car. Again, these are decisions that only you can make. Remember that if you are not trading in a car with equity in it (if the gross trade allowance, or GTA, is *less* than the closeout balance, that is considered *minus-equity*), the lender may insist on a payment, depending on the personal strength of your credit. A down payment can decrease your monthly payment, increase the price of the car within your budget, or decrease the term of financing. (*Important!* If a *factory-to-customer* rebate is being offered on the vehicle that you ultimately decide to purchase, the rebate should be added to the down payment.) Enter the down payment on your worksheet.

7. ACV of trade-in (if applicable). Refer to Chapter Six, To Trade or Not to Trade, and follow the instructions for establishing the actual cash value of your trade-in. Enter the ACV on your worksheet.

8. Closeout on trade-in. If you owe money on your trade-in, call the lender and verify your closeout balance. Closeout balances are always qualified with a specific date. For example, if you called your lender on March 1, you would probably be told that the balance quoted was valid until March 15. This simply means that if you should pay the closeout or purchase a new car *after* that date, you will need to get an updated closeout balance. I suggest that you have your account number readily available when you call the lender to help expedite your request. Enter the closeout balance on the worksheet.

Before we calculate the maximum amount you can afford, given all of the disclosed information, let me caution you. It is entirely possible that the car you wish to purchase may be too expensive for your overall budget. If you have set your sights too high, you may have to compromise somewhat and consider your second or third choice simply based on economics. Don't let this discourage you. We are all victims at one time or another of having champagne taste on a beer budget. A $20,000 car is not twice as good as a $10,000 car. It simply has more bells and whistles.

You should be happy to know that most of the best-performing cars, with the highest resale values and the most favorable histories, have been the modestly priced ones. So do not feel that you are compromising your goals just because you may not be able to afford the car that is Number 1 on your wish list. Besides, unlike a decade ago, there are so many high-quality cars to choose from today that your most challenging task may be narrowing the field.

Calculating Your Budget: Now I will show you how to use all of the information you have just entered on the Budget and Payment Worksheet. It may sound a little confusing at first, but I will clarify each calculation and explain the objective. To simplify this process, I will use the same hypothetical numbers used earlier. I have not included registration and title fees, which are usually incidental amounts. It is assumed that these fees will be paid out of pocket. Refer to the Sample Budget and Payment Worksheet in the Appendix, page 00.

First, we need to establish how your desired monthly payment translates into real dollars financed. The monthly payment factor you entered on the worksheet represents the desired term in months and the projected interest rate. Divide the desired monthly payment by the factor:

$$\$250 \text{ divided by } .02658 = \$9,405.57$$

This simply means that if you finance \$9,405.57 for 48 months at an APR of 12.5 percent, your payment will be \$250 per month. Add the down payment to this figure:

$$
\begin{array}{r}
\$\ 9,405.57 \\
+1,200.00 \\
\hline
\$10,605.57
\end{array}
$$

Subtract your closeout balance:

$$
\begin{array}{r}
\$10,605.57 \\
-750.00 \\
\hline
\$\ 9,855.57
\end{array}
$$

This gives you your total equity, including sales tax but excluding the value of your trade-in.

Remove the sales tax from the total equity:

$$\$9,855.57 \text{ divided by } 1.07 = \$9,210.81$$

This is a *very* significant figure in the overall calculation. \$9,210.81 is the *maximum cash difference* you can pay to achieve a monthly payment of \$250. In other words, let us assume that the car you are interested in has a suggested retail price of \$12,500. You would have to achieve a combination of discount and/or trade-in equal to the difference between \$12,500 and \$9,210.81.

Retail Price	$12,500.00
Trade-In	−3,289.19
Cash Difference	$ 9,210.81

Confused? Let me restate a couple of key points. Remember that this calculation was made from specific, predetermined numbers entered on the Budget and Payment Worksheet. If you do not have a down payment, a closeout, or a trade-in, those calculations would be removed from the computation. Remember also that one of your ultimate goals is to enter the negotiating arena fully armed. This means that you not only need to know the specific car you are interested in purchasing, the dealer invoice on that car, and the actual cash value of your trade-in, but it is imperative for you to know in advance your budget limitations. This vital advance knowledge will ensure that no matter how cunning the salesperson, you will not be bullied into buying a vehicle that you simply cannot afford.

We have determined that $9,210.81 is the maximum cash difference you can afford to pay for any car. Now we can go a step further and determine the most expensive car you can afford. Add the maximum cash difference to the ACV of your trade-in:

$$\begin{aligned} & \$\ 9,210.81 \\ +\ & 2,000.00 \\ \hline & \$11,210.81 \end{aligned}$$

It is my contention that 95 percent of all cars sold in the United States, domestic or foreign, that have a *sticker price of $30,000 or under, can be purchased for somewhere between dealer invoice and $500 over invoice.* The only factor that will determine this variable is the individual negotiating ability of each buyer. With this in mind, we should play it safe and allow the maximum dealer profit of $500 to determine the highest dealer invoice car that you can afford. If you negotiate a deal that is better than $500 over invoice, then your monthly payment will be a little lower than anticipated.

If we take $11,210.81, which does not allow for dealer profit, and subtract a projected $500 maximum profit:

$$\begin{array}{r} \$11,210.81 \\ -500.00 \\ \hline \$10,710.81 \end{array}$$

Pay attention! Factoring in all of the information from the Budget and Payment Worksheet, we have now determined that the dealer invoice price, *including* all factory-installed options and maximum dealer profit, cannot exceed $10,710.81. In other words, you can negotiate a deal on any domestic or import car you choose as long as the total dealer invoice does not exceed $10,710.81, and as long as the cash difference you pay does not exceed $9,210.81. These two *very important numbers* will enable you to achieve your monthly payment goal of $250. If you do not have a trade-in, then the total maximum dealer invoice cannot exceed $9,210.81 minus $500, or $8,710.81.

You are going to have to play with these figures a little bit to make them work and to ensure that the concept is clear in your mind. Practice with some hypothetical numbers just to be confident that you fully understand the mathematics involved.

Now that we have determined how much you can afford, we must go a step further. Let us consider the other important factors that will enable you to intelligently choose the *most* car for the *least* money.

Functions and Features

The second thing to consider before selecting a specific car is functions and features. Do you need a four-door, automatic transmission, air conditioning? Are miles per gallon important? Do you need a four-, five- or six-passenger car? Is a station wagon ideal? How about headroom, color, overall aesthetics, performance? Do you need a hatchback, a big trunk, a four-wheel drive? All of these things should be considered.

Sit down with a piece of paper and make two lists. The left side of the paper is titled *Must-Have Features*. Start listing those options and functions that are essential in your consideration of any model car. The right side of the paper is titled *Wish List*. These are the items that you would really like to have, but you can live without if they put you over your budget. Only you can judge which options are must-haves, and which are wishes.

Now you can price out a car exactly the way you want it. If it is over your budget (and most of the time it will be) you can start eliminating options by virtue of their importance to you. Selecting a car within your budget is a fundamental rule in achieving the best possible deal.

You will need to use the buyers' guide that I instructed you to purchase. Look up dealer invoice on any models that you are considering. Add options, subtract options. Play with the numbers a little bit. Get a feel for the price range you can afford. You do not have to make any earth-shattering decisions at this time, but it would be helpful if you start to establish some ranges.

Frequency of Repairs

Car buyers tend to forget that a new car doesn't stay new forever. When a car reaches the point where the original factory warranty runs out, the hard, cold reality of life is that a car that has consistently shown a poor history of dependability can bleed every penny out of your savings account. Service labor rates of thirty, forty, or even fifty dollars per hour, along with outrageous prices for parts, can make your dream car rapidly turn into your biggest nightmare.

There are dozens of publications that clearly outline the track record of virtually every car in the world. Do some research. Spend twenty dollars on a half-dozen consumer magazines that show detailed histories of every model car sold in the USA. It can save you thousands of dollars in repair costs. Most of these magazines objectively evaluate frequency of repairs and dependability factors of every manufacturer and every model. *Edmund's Used Car Ratings* is one publication that is loaded with helpful information on how various cars have fared and their history of dependability. *Consumer Reports* also publishes a book devoted to the history of repairs and overall dependability comparisons for all domestic and import cars and trucks.

The key to this research project is not necessarily to select the exact model car you should consider, but to eliminate the poor choices. Try to narrow the field. This will make the choosing process much easier. Once you have eliminated the poor choices, you can use other factors to consider your ultimate decision.

It is time for you to be open minded. What you have purchased in the past, as well as blind loyalties that have victimized your pocketbook, must be discarded—no more preconceived ideas that will prevent you from making an objective, intelligent choice.

Resale Value

Except for those of you who buy a new car and drive it until it dies, resale value may be the most important factor to carefully evaluate when narrowing your selection of cars. I cannot impress upon you enough the magnitude of this point. Resale value never seems to be an important consideration when people shop for a new car. People tend to forget that the car you buy today is the car you trade in a couple of years. The trade-in value of your car has a significant effect in determining the ultimate cost of a new car.

You might be shocked to learn that some cars

depreciate so dramatically that an assumed savings at the time of purchase can equate to a loss of $4,000 or $5,000 at trade-in time. And to complicate things even further, there are some very good cars that unfortunately have a history of poor resale value. Why? Supply and demand, for one. That a particular car meets all of your requirements at the time of purchase does not guarantee that it will hold its value.

There are many methods of establishing the projected depreciation of a car. Unfortunately, because none of us has a crystal ball and can predict the future, there is no foolproof method. But there is a best method.

The first thing I would like to do is to use the term residual rather than resale value. Residual is a term used by banks and leasing companies. It simply means *expected* future value. Whenever a car buyer leases a car rather than buys one, the bank determines the residual value of the car. The future value is significant in determining the monthly payment. The fact that most banks and leasing companies spend thousands of dollars on research to accurately determine the residual value of every car and every model imaginable makes them the best possible source for estimated future values of cars.

Banks and leasing companies are literally renting vehicles to consumers. A lease is a rental and the leasing company owns the car at the end of the lease term. Rest assured that the residual values used by banks and leasing companies are substantiated by dozens of reports and an enormous amount of research. Leasing companies cannot risk unrealistic future values on the cars that they lease. If the residuals they use are not accurate, they can be hurt in two ways.

If the residual values are too low, they will be unable to compete with other leasing companies. If the residuals are too high, they risk losing all of the profit from the lease when the car is sold at the end of the lease term, for less than the projected future value. It is a tricky business.

Once you have narrowed your selection of possible new cars down to three or four models, using the other criteria I have given you in this book, you can compare residual values to help with the ultimate decision. Where exactly do you get residual values?

First of all, you can look in your phone book yellow pages under leasing. Call several leasing companies and tell them that you're considering leasing a new car. But you want to lease a car that has a high residual value to keep your payments in line. Ask them for the residual values on the cars that you are considering. Get the residual values based on 48- and 60-month terms. Compare the values to help determine what car will be your best buy. You can also do the same thing with any bank that has a leasing department.

Now please understand that residual values and ultimate resale values can be considerably different. No one can accurately predict what market conditions will be like in five years. The residual value of a vehicle does not suggest that that is what the vehicle will sell for in four or five years. However, comparing one residual value to another gives you a credible barometer to establish what cars you should consider, based solely on their relative future values. It is the best way to compare apples with apples.

To put all of this residual business into practical terms, let's crunch some hypothetical numbers. Assume that you are considering three different cars. Although each car has a sticker price of $12,000, the residual values will be 25 percent, 35 percent and 45 percent at the end of a 48-month lease. It may not seem logical to you that a variation of 20 percent is possible on two cars that have the same sticker price, but in some instances it can be even higher.

	Sticker Price	Residual %	Dollars
Model A	$12,000	25%	$3,000
Model B	$12,000	35%	4,200
Model C	$12,000	45%	5,400

To put this exercise into even more realistic terms, let's also assume that you negotiate your best possible deal on each of these three cars. Model A can be purchased for $11,000; Model B for $11,500; and Model C is so much in demand that sticker price is the best you can do. Now if you do some simple mathematical comparisons, it is easy to see that even though Model A initially cost $1,000 *less* than Model C, what happens at trade-in time? You guessed it. The resale value on Model C is so much better than Model A that the $1,000 savings at the time of purchase turns into a $1,400 *loss* at trade-in time!

Bear in mind that residual values are projected

and there are a dozen variables that will ultimately affect the true resale value of the car you choose. However, all things being equal, I am unaware of any other method to establish future resale values on automobiles as accurate as the residuals used by banks and leasing companies.

With all of the information in this chapter, your buyer's guide, and other consumer repair magazines, you should be really zeroing in on three or four specific model cars that meet all of your criteria and will ensure dependable service and an above-average resale value. The next chapter brings you face to face with the salesperson—not to negotiate, but to gather vital information. Read on!

10

Formulating the Plan

Now that the prior chapters have removed some common myths and misconceptions, established a budget, and helped you value your trade-in, it is time to come face to face with the salesperson—not to negotiate a deal on a new car, but to gather the information you'll need to complete the step-by-step process to get the ultimate new-car deal. Before you even venture out your front door, you will be thousands of dollars ahead of the pack just because you have learned that the price you pay for a new car is only part of the picture. Hopefully, if you follow my advice, you will carefully consider all of the criteria disclosed in this book to intelligently select a car that will give you years of trouble-free service and hold its value.

Based on the information provided in prior chapters, you should be able to narrow your possible selections down to three or four models. It is necessary for you to consider dealer cost on the models you are interested in and the actual cash value of your trade-in. Although you are not going to actually negotiate at this time, the three or four cars that you are considering *must* fall within the limits of the Budget and Payment Worksheet. Needless to say, you must shop for models that meet your budget guidelines. It is important to understand that this particular part of the car-buying process is vital. At this juncture you will be gathering essential information that will allow you to decide on a specific model car and a specific dealer.

Before you can make an intelligent decision on selecting a new car, you must obtain specific information directly from the dealer. This information cannot be found in any book. Your mission is not to talk price, but to decide on a specific model and a specific dealer. The best price can be negotiated at any dealer that you choose. First, there is other information that you will seek to assist you in making a buying decision.

The basis for selecting a specific model car will depend on your ability to objectively evaluate all of the information that you gather. It is a very personal decision. The logic for selecting a dealer will be determined by your opinion of which dealer offers the most exclusive overall benefits to you. Forget price. Remember that the lowest price can be negotiated at any dealer you choose. Price is not the issue at this point.

You can shop as many dealers as you like until you feel comfortable that your choice of model and dealer is justified. Try to gather your information on a weekday, if possible. Saturdays, Sundays, and evenings are usually busier times for dealers and it may be difficult to find a salesperson who will take the time to answer all of your

pr...
will be un...

When you enter... salesperson, the first five min... is vital. The salesperson will try to size y... "qualify" you. He will be trying to determine... you are a today buyer, a "flake" (a professional shopper who has nothing better to do than waste a salesperson's time), or a serious prospect who just isn't quite ready to make a buying decision. What you say and how you say it is critical. You must let the salesperson know in no uncertain terms that you are a serious buyer, you are not wasting his time, and that you intend to buy a car within a couple of days.

It is extremely important for you to be right up front with the salesperson. This is not the time for games or masquerades. Tell him *exactly* what you want and what your intentions are. This is what you say. "Hello, my name is Janet Smith. I would like some information on an XYZ automobile. I *am not* prepared to make a deal today. However, based on the information that you provide, I may be interested in working some figures in the next few days." Make your position crystal clear by adding: "I must warn you that I am not interested in buying today. If any attempt is made to sell me a car today, I will immediately leave and take my business someplace else. Please *do not* ask me what you can do to sell me a car today."

e and nega-
on your overall
de

Ta _____ ry sticker and make note
of the _____ s suggested retail price and
the des _____ harge. Also be sure that the majority of _____ factory-installed options are the ones
you are interested in. Note the retail prices of
them as well. If there is a supplemental sticker
on the car that outlines after-market options or
accessories, ignore it. Now it is time to sit down
in the salesperson's office and ask him a series of
questions. Your questions should go something
like this:

"I am considering buying an ABC, an XYZ, or
a JKL. I haven't quite made up my mind yet. Tell
me why the XYZ is a better choice than the other
two models I am considering." Ignore any reference made to price, discounts, or trade-in allowance—these factors are meaningless. Any sales
sizzle that the salesperson tries to lay on you
should be taken with a grain of salt. You are inter-

ested in features and benefits only. Be sure to distinguish between sales sizzle and facts. Make notes of the facts and their benefit to you. After the salesperson has completed a competitive analysis, ask these additional questions:

1. What is your service labor rate per hour?
2. Do you provide service-loaner cars? If so, are they free of charge?
3. Do you have a customer service shuttle bus?
4. What are your service department hours? (Find out if they are open evenings or Saturdays.)
5. Do you have a customer relations manager? If so, what is his or her name?
6. Do you have an accessory menu? (An accessory menu is a catalog or brochure that lists all of the dealer installed accessories available and the retail price of each.)
7. What is your CSI rating? (CSI stands for Customer Satisfaction Index. Every car dealer across the country is rated by the manufacturer based on owner surveys called CSIs. Ask the salesperson to *show* you the dealer's latest CSI rating. If he gives you a song and dance, it could mean that a problem exists with the rating. Be insistent. Tell him that the dealer's overall CSI rating will be a strong factor in determining with whom you ultimately do business. If he insists that he cannot show you the survey, strongly consider eliminating this particular dealer from your choice.

Bear in mind that the CSI rating is a dealer's report card and that it is graded by past customers. A dealer's reluctance to expose a report card can only mean that it is a poor one.)

8. Can you give me any unique or exclusive benefits that I would realize by doing business with you rather than with one of your competitors? (This is the salesperson's moment of truth. It is his opportunity to hit you with every trump card in his arsenal of sales sizzle to ensure that you will return and buy a car from him. Remember to record only meaningful facts, things that are benefits to *you*.) If he makes reference to price, discount, or trade-in allowance, do not even bother to comment. *You* will control the price, discount, and trade-in.

9. Are there currently any factory rebates or reduced interest rates available on the model(s) that interest me? Ask the salesman to show you in writing the specific guidelines of any available incentive.

10. There may be other specific questions that you wish to ask. Do so at this time. Do not lose sight of your goal. Gather information that will enable you to make an intelligent buying decision. Thank the salesperson for his time and tell him that you will be in touch with him soon. The salesperson may try to get some information from you at this time or try to get you to reconsider your decision not to buy a car today. The sales manager may even try to sell you. No

matter what they say, leave as soon as you have gathered the information that you were seeking.

Do you have any idea what a mind-boggling impact that entire approach has on the salesperson? How often do you think the typical salesperson encounters a customer who spends about an hour with him and never even asks for a price? Take it from me, you have not only accomplished your intended goal of gathering information, but you have also established a position of control.

He hasn't a clue why you didn't ask him for his best price. After all, every other customer in the world hammers him to death for the elusive best price. This is a noteworthy victory for you. The salesperson never had the opportunity to take control. You set the pace and kept him on the defensive.

When you ultimately return to a particular dealer to begin the negotiating process, as soon as the salesperson sees you, he will realize that you are not the typical, naive car buyer. He will know that he has his work cut out for him if he hopes to sell you a car. He will also come to the realization that gimmicks and double talk will be totally ineffective. Victory is just around the corner!

Talk to as many dealers as possible, until you are confident that you have enough information to decide on a specific car and a specific dealer. You are now just moments away from sitting across from any car dealer and winning the most exhilarating victory of your car-buying existence!

11

Winning at the Negotiating Table

Congratulations! You are just moments away from sitting across from *any* car dealer in the country and beating him at his own game. Before you enter the negotiating arena you must take some time to assemble your plan. You must organize your information and make some decisions. You must take out your calculator and crunch numbers. Remember, when you enter the dealer's showroom to negotiate a deal, you cannot leave *anything* to chance; plan your work and work your plan.

Before we get into specifics, I must make several very important points. We must set the overall tone of your negotiating strategy. First of all, one of your goals during the entire negotiating process is to be as emotionless as humanly possible. Your demeanor should remain almost robotic, no matter what is said or how the dealer tries to intimidate you. One of your strongest weapons is to confuse the salesperson. If he can't read you, he can't beat you.

Secondly, do not get into any debates with the salesperson, the sales manager, or a closer during the negotiating process. They will do everything possible to unravel you and challenge your logic for making the offer you are going to make. Do not justify your position. Do not try to reason with them. Be polite and professional, but remain as "dumb as a fox."

The only explanation for your offer is this: You know how much you can afford and you do not wish to compromise the specific car or options you have chosen. *If* you can buy the car for the cash difference offer you have made, they will sell a car and you will be a happy customer. If they cannot meet your price, you will try to achieve the price from a competitor. Should no one be able to meet your price, then you are simply going to postpone your purchase.

Please be Careful! Do not allow them to talk you into buying the car for a specific payment rather than a cash difference—the old "if-I-could-would-you" routine. Convincing car buyers to focus on a monthly payment rather than a cash difference has always been an effective tactic used by car dealers. Think about it. If they can focus your attention on a monthly payment as the only criteria for you to evaluate a deal, then your entire mission—all the planning and number crunching—will have been in vain.

Psychologically, it is much easier to convince a customer to accept a payment that is $6 a week more than she has offered than it is to convince her to pay $1,000 more. On a 48-month loan at a 12 percent APR a $6-a-week increase in payments translates to over $1,000! See how easy it is to get tricked? When negotiating a deal on a new car, always make the cash difference the issue. If you even hint that your objective is to achieve a specific monthly payment, you will seriously compromise your negotiating edge. Only *you* know that your offer is based on a monthly payment. Remember—*dumb like a fox*.

Salespeople and sales managers are as human and emotional as car buyers (perhaps even *more* emotional). Your attitude toward them is an important tool during negotiating. These people are high strung and they live in a daily environment bursting with pressure and tension. Dealers have been known to cut off their noses to spite their faces.

I have seen dozens of well-informed, shrewd car buyers negotiate an outstanding price on a new car and then blow the whole deal just because they displayed such a pompous, obnoxious, holier-than-thou attitude. The sales manager refused the deal on the basis of principle! It's true. A sales manager will sometimes refuse a deal he would normally accept, totally based on a dislike of a customer whose attitude was offensive.

That is why your attitude should reflect a decent, honest, friendly customer—just an average car buyer. Even if you have to give an Academy Award performance, display an attitude that will make the salesperson like you. Day in and day out, the average salesperson is bombarded with rude, sarcastic car buyers who make their job really tough. If you are a breath of fresh air, you will be amazed at how hard the salesperson will work for you.

At this point, you should be in a position to select a specific model car and a specific dealer. You know how much you can afford and you know dealer invoice on the car you wish to purchase. Remember that the best time to negotiate a deal is at the end of the month. Be sure that you do not walk into a dealership on *the* last day of the month. If you do, the end-of-the-month theory will not help you. The dealer needs to have a couple of days to prepare paperwork, prep your car, and arrange financing before actual delivery takes place. Your purchase has to be included in the end-of-the-month business in order for it to be an advantage to the dealer. If the sale is too close to the end of the month, it may drift over into the next month's business.

Do not ever lose sight of the undeniable fact that buying a new car is the purest form of horse trading in the world. There are no shortcuts. If this book has taught you anything, it has taught you that the best price is *totally* determined by your ability to negotiate and your preparedness when you enter the showroom. Also remember that the process of negotiating has to include dickering. A car dealer will not accept *any* deal until he is certain that he has squeezed every penny out of your pocket. Only then will he succumb to your stamina. It is a game of cat and mouse and there is simply no way around it. Those car buyers who have tried to beat the system by avoiding the negotiating process have walked away from the car dealer unfulfilled.

The negotiating process is the ultimate example of psychological warfare. Remember, you are fighting the car dealer with your brain, not your sword. Make up your mind right now that you will have to play the back-and-forth game with the salesperson and sales manager. Trust me, it will be unlike any prior experience. *You* are in control and *you* will win the prize. Besides, won't it be just a little bit of fun turning the tables on the car dealer for a change? Instead of the dealer making the decisions, you will be in control of your own destiny. Your research, planning, and number crunching is about to pay off. You have left nothing to chance. You know exactly what your budget guidelines are going to be, the value of your trade-in, and how to achieve a comfortable monthly payment. You have the assurance that the car you selected is a quality model with a strong resale value and an excellent history of repairs.

This brings us to a very controversial point in this book. You now know how much you can afford to spend, how long you want to finance, the ACV of your trade-in, the closeout balance, and the dealer invoice on the car you wish to purchase. But the main question here is this: How much profit should the dealer be entitled to make? What is fair and what is an exceptional deal for you?

I could have devoted an entire book to such a debatable topic. However, let me capsule it for you. It is obviously your goal to purchase a new car for the lowest possible price. I sincerely hope that this book has opened your eyes to the fact that the lowest price is only a small part of the overall strategy used to achieve an outstanding deal. As we have discussed in great detail, many other factors truly determine the *real* cost of ownership from date of purchase to date of trade-in. At the risk of entering into a philosophical explanation, defining the lowest price needs some clarification.

There are those who have written books such as mine, who would lead you to believe that you can buy a Chevrolet Cavalier, a Honda Accord, or a Jeep Cherokee for a carved-in-granite specific price, over invoice, on any given day. To arbitrarily make this claim is misleading and dangerous. I wish that I had the power and insight to do this, but it is simply impossible. It is beyond comprehension or logic for anyone to accurately predict the exact price a specific car can be purchased for, on any moment of any day. There are far too many variables and market factors that affect one's ability to achieve a specific price on a specific car.

Therefore, the only logical way to find this elusive lowest price is to establish guidelines and ranges. One must base these guidelines on history, viable research, and full knowledge of the intricacies of the automobile industry. I must warn you that due to an ever-changing car market and factors beyond anyone's control, each negotiated deal is exclusive. Every dealer is unique and each car buyer different.

The fundamental theme of this book is to ensure that anyone who follows the guidelines and suggestions outlined will not only buy any car on any day for an exceptionally low price but will benefit during the entire period of ownership. I assure you that you will be gratified as never before when you enter the negotiating arena with a car dealer and consummate an outstanding deal.

As I have stated earlier, it is my conviction that with few exceptions, any new car, domestic or import, with a retail price of under $30,000, can be purchased for somewhere between invoice and $500 over invoice. You are in control of your own destiny. If you do your homework and leave nothing to chance, the deal you ultimately negotiate should be closer to dealer invoice than to $500 over invoice. Your fortitude and stamina will determine the future of your negotiating ability.

Now that all of our homework is done, let us plan your negotiating strategy. Review your Budget and Payment Worksheet and follow the guidelines herein. For the sake of clarity, I will use the numbers from the sample worksheet in the Appendix. We know that your maximum monthly payment is $250. By giving yourself a $500 range for negotiating, you will be able to make your first offer low; subsequent offers will not jeopardize your budget decisions.

It is imperative for you to fully understand the math and the relationship between three key numbers. First of all, looking at the Sample Budget and Payment Worksheet, note that (4) is $9,210.81. This figure is the maximum cash difference you can pay for any car, considering all given variables, and still achieve a payment of $250 a month. Secondly, the next important number is (1) on the worksheet, which is $9,405.57. This is the maximum amount you can finance to ensure a payment not to exceed $250 per month. Last but not least is (6) on the worksheet, $10,710.81. This number reflects the highest dealer invoice car that you can consider to maintain your payment objective. Let me now explain to you how these all important numbers are used during the negotiating process. You should be prepared to devote at least an hour strictly to negotiating. From the moment you make your first offer it is essential that the dealer recognizes that you are a serious buyer, not a pushover. Remember that you are going to be dumb like a fox and you are not going to get involved in any pointless debates.

Make absolutely sure that you have done all of your homework. You have to know what your maximum budget is. You must know the ACV of your trade. The closeout on your trade-in should be fresh on your mind and you should have a good feel for interest rates. Remember that if you are going to purchase any dealer-installed accessories or an extended warranty, all of these factors must be used to determine your overall budget.

Call the dealership you have chosen and ask for the specific salesperson with whom you spoke on your information-gathering visit. When you call him, make your intentions clear. Tell him that you would like to set up an appointment to come in to the dealership to *buy a new car*. Notice that I did not say to negotiate a deal on a new car. You want the salesperson to know that you are returning to the dealership to buy a car. This will assure you that his adrenaline will be flowing when you enter the showroom and he will not get involved with another customer prior to your appointment.

Do not bring any buyer's guides or any publications with you that would indicate that you are an educated buyer. A small calculator is okay, but you want to be sure that the salesman and sales manager assume that you are just an average customer. Be punctual. You should arrive at the dealership about fifteen minutes prior to your appointment. When you enter the showroom the salesman will be very gracious indeed. You see, he spends most of his time talking to customer after customer who continues to beat on him for his best price. It is always a refreshing change when the salesman knows right up front that he will be working with a "today" buyer.

After the usual small talk, restate your mission. Tell him that you would like to buy a car today, but that you would like to drive the specific model you will be purchasing. If he does not have a car in stock that is exactly what you want, that is okay. But be sure he does not try to sell you a model with more options than you desire. If it has to be ordered or traded from another dealer to get *exactly* what you want, insist that he do so.

Let him demo the car one more time so you can be certain that you have made the right choice. If there are any questions about the car or the specifications, get this out of the way during the demo ride. When you sit down at the negotiating table, the only thing you want to concentrate on is the deal. After the demo ride, look the car over one more time and if everything is all-systems-go, tell the salesman that you would like to make him an offer on the car. He will be noticeably excited.

Remember that you have three important numbers implanted in your brain. The maximum cash difference computed on the Budget and Payment Worksheet, the maximum amount to finance to achieve your affordable monthly payment, and the retail price of the model you are purchasing. Be certain that the model you are buying has a manufacturer's suggested retail price that matches the price disclosed in your buyer's guide, including all factory-installed options and destination charges. If you have decided to purchase any dealer-installed accessories from the accessory menu that you picked up after your initial visit, do not include them in negotiating on the car itself. Even if you have to pay for dealer-installed accessories with a Visa or Mastercard, *do not* include accessories in the negotiating process. Be sure before you start negotiating that the price the salesman writes on the buyer's order matches the sticker price and the option prices you have calculated from your buyer's guide. This is not the time to be comparing apples to oranges!

As I have stated and restated throughout this book, buying a car in the Nineties is a game that can only be won at the chess table, not on the battlefield. I cannot impress upon you enough the importance of your demeanor and the manner in which you negotiate. No matter what the salesperson says to try to dilute your strategy, you must not waver. If the sales manager or a closer tries to put the hard close on you and really becomes a little overpowering, it would be better to leave than to subject yourself to this kind of pressure.

Please bear in mind that as I walk you through the negotiating process, it would be impossible for me to outline a specific script that is all things to all people. Each deal that is negotiated involves diverse conditions and unique personalities. You will have to use a little intuitiveness and improvisation as the negotiating evolves. The scenario that follows is a general plan of attack.

Now that you are sitting in the salesman's office and have concurred with the retail price and factory options to be added to your car, it is time to dig in and go to work. Offer the salesman $1,000 *less* than the maximum cash difference you computed on the Budget and Payment Worksheet. Again for clarity, I will use the numbers from the Sample Worksheet. Your first offer will be $9,210.81 minus $1,000, or $8,210 (drop the 81 cents change—the salesman would really get suspicious if your offer was $8,210.*81*).

Offering the dealer $1,000 less than your maximum cash difference gives you a negotiating edge. Believe it or not, *most* car buyers make offers that are $3,000 to $4,000 under invoice! Your *reasonable* offer makes your deal obtainable in the dealer's eyes.

If he is like most salesmen, he will almost immediately try to convince you that your offer is much too low. Let him give you his sales pitch without arguing or debating with him. Give him a check for $500 (if you cannot afford $500, then $100 is almost as effective) to convince him that you are not playing any games. Submitting a check with an offer is the best way that I know to get a sales manager's attention. Also, sign the offer to further reinforce your honorable intentions.

If you have a trade-in, the salesman will get it appraised at this time and will bring the appraisal on your car, your signed offer, and the $500 check to the sales manager. After making you wait for a little while (which is part of the dealer's strategy) the salesman will return and say one of two things. Either he will return with a ridiculously high counter-offer or he will simply say that your offer is much too low and that you will have to "get your thinking up."

Let him give you his sob story, listen as if you really cared, and when he is done offer him an additional $200. Your original offer of $8,210 is now $8,410. More than likely he will tell you that an additional $200 just is not enough. Be firm, yet polite, and insist that he take your counter-offer into the sales manager. He will again make you wait for a while and return with basically the same story and possibly a different counter-offer than the sales manager's original counter.

After the salesman tries to convince you of how much money he will lose if he sells you a car for $8,410, play with your calculator for a couple of minutes and offer him an additional $150. This now brings your offer to $8560. He will again talk to the sales manager and return with bad news. He will probably sigh and repeat every argument in his handy-dandy salesman's guide, but be polite, let him speak his piece, and do not debate. Tell the salesman that you need a couple of minutes to think. If you are with your spouse or a friend, simply say that you would like to discuss any further offer with your friend, husband, or other companion.

Get up from the salesman's office and take a little walk. Walk around the showroom; take a walk outside. Kill a little time and let the salesman and sales manager sweat a little. If you are with another person, give the appearance that you are having a deep conversation. Return to the salesman's office and say, "This is my final offer. I will pay you $8,710, plus tax and license, and that's as far as I am going to go. If the sales manager does not accept this offer, please return my good-faith deposit and I will leave."

If the dealer accepts this offer (and he probably will not, unless he is a pushover) you have just bought a car for right around invoice. This is a very critical crossroad in the negotiating process. It is time to play hardball and may the best player win! More than likely, the salesman will return to his office not with your check, but with a closer or the sales manager. Rest assured that whoever returns with the salesman will be a seasoned professional who specializes in extracting money from customers who insist that they will not pay one penny more.

These specialists will try to convert you to payments. They will plead with you to give them just "a little help." You will be astonished at the effectiveness of their arguments. Do not deviate from your game plan. Listen to his sales pitch and insist that you are not going to pay one penny more until you have had a chance to check out another dealer. (If you should know the name of this dealer's biggest competitor, use their name specifically.)

After he exercises his last-ditch effort to persuade you to pay more, throw him a bone. Tell him that you are really tired of shopping around and that just to show your good intentions, you will give him $100 more. This will bring your offer to $8,810. Remember, you *cannot* exceed $9,210.81 to ensure that you do not surpass your budget. You may get lucky. If it is close to the end of the month and he really needs the sale, he may take the deal. If he does, you have just negotiated a deal for about $100 over invoice. If not, you will be forced to grandstand. Tell the salesman or closer to take another look at your trade-in. Surely they can find a way to squeeze enough money out of your trade to consummate a deal. If you are at an impasse, stand your ground.

Ask the salesman and closer for your deposit check back. When they return it to you, thank them for their time and tell them that maybe the next time around they can earn your business. No matter what they say, except of course, that they accept your last offer, you should make your intentions crystal clear. You are going to go home now, but tomorrow you are going to check out their competition. Tell them to call you if they change their minds about your offer, but politely warn them that they had better not wait too long.

How much farther should you go? I believe at this point that you should actually leave the dealership. Make them understand in no uncertain terms that you mean business. Believe me, they will call you. If they let you leave without yielding to your last offer, you will probably be forced to pay a couple of hundred dollars more when you return. Your final offer was $400 below the maximum cash difference within your budget. You may want to try a competitor. It is up to you. Just be sure that you realize that it is the end of the month and *all* car dealers fight and grapple for every deal. The competition between dealers is so ferocious that a dealer would have to be crazy to let you leave only a few dollars away from a deal. I assure you that if you negotiate as I have suggested, you will not pay more than $500 over invoice for any car under $30,000.

Does all of this sound too simple to you? Does it appear that buying a car and negotiating is a lot easier than you thought or remembered? Can this be too good to be true? Don't kid yourself. The only reason that it seems to be like shooting fish in a barrel is because for the first time in your life, you entered the negotiating arena *totally prepared*. All the research, number crunching, and planning has just come to fruition. Give yourself a pat on the back and take pride in the fact that you have accomplished a feat that few car buyers will ever achieve.

After you have made your deal, look at the numbers on the final retail buyer's order *very carefully* before you sign it. Do not take anything for granted. Check the retail price, the cash difference, the license fees, the closeout balance on your trade-in, and your cash on delivery, and be sure you receive a signed copy of the retail buyer's order with all of the figures disclosed.

When the salesman finishes with the paperwork,

he will probably turn you over to the finance and insurance (F & I) manager. This guy will try to convince you to finance through the dealer. If his rates are competitive with local banks, it would be convenient to use his financing. A word of caution: Before signing the bank contract, you must also scrutinize all of these numbers as well.

The F & I manager may try to sell you after-market accessories or an extended-service agreement. Refer to Chapter Four, Misconceptions, Myths, and Fairy Tales, and follow my instructions on these subjects.

When the time comes for you to take delivery of your new car, again you should scrutinize *everything*. Check the bank contract very carefully. Look at the APR, the amount financed, the monthly payment, the term of the loan, and so forth. Do not assume that the great deal you negotiated cannot be altered with a little creative math. Also, look your car over very carefully. If at all possible, try to take delivery during the day, so you can check the car from top to bottom for paint defects or anything that is not to your liking. Once you leave the dealership, it is very difficult to get an adjustment on a paint chip or a scrape on a bumper.

Make certain that you have the warranty book, maintenance schedule, and tire warranty on your car. Record the ignition and door key numbers in a safe place in case you ever lose your keys. Ask the salesperson to give you a tour of the dealership. Have him give you the name of the general manager, the service manager, and the parts manager. If there is a customer relations manager, ask to be introduced to him or her.

Congratulations! You are the proud owner of a sparkling new car that you purchased for a very competitive price. You negotiated an excellent trade-in allowance and a low interest rate, and you have the security of knowing that the car you have chosen will give you years of trouble-free service and maintain an excellent resale value.

12

A Word on Leasing
a Car

For those of you who may consider leasing a car
rather than buying one, I will try to remove some
of the mystery. Although the car-buying process
has many facets, most of which are misunderstood, leasing seems to win the award for most
complicated of all. Leasing has existed for many
years, but consumer leases have become popular
only recently.

There are many different ways for a leasing
company or car dealer to structure a lease, but I
will focus on the most popular methods. I will also
warn you of the pitfalls and things to carefully consider. Remember that leasing laws are governed
by individual states and that guidelines for leasing
companies may vary considerably from one state
to another. If you have any questions about the
legal language of any lease contract, consult an
expert before entering into an agreement.

A lease is a fancy way of saying rental. Leasing
a car in reality is renting it for a predetermined

amount of time at a predetermined monthly rental charge. The leasing company (the lessor) owns the vehicle and you (the lessee) have no equity interest. Most cars can be leased for two, three, four, or five years.

There are five factors that determine exactly how much your monthly rental charge will be. They are: the term of the lease, the capitalized cost of the new vehicle (selling price), the residual or future value of the car, the interest rate or lease factor used to calculate the lease, and the number of miles driven during the lease term. Since you have no vested ownership in a leased car, your main goal will be to lease the *most* car for the *lowest* monthly payment.

The term of a lease should be carefully considered. Obviously, the longer the term, the lower the payment, but there *is* a catch. Unlike a retail installment contract, where you own the car, you do not have the luxury of terminating a lease before the term is up without incurring severe penalties. Be prepared to drive the leased car for the entire lease term and make your decision with that in mind.

The capitalized cost is the amount the car dealer is charging the leasing company for the car. In effect, *it is the price that* you *are paying for the car*. A word of caution: The leasing company could not care less whether or not you get a good deal. If the dealer can talk you into leasing a particular car for $250 a month and that translates to selling the car to the leasing company for $2000 *over* factory sticker, you are the one who pays dearly, not the leasing company. Most leasing companies will allow a dealer to lease a car for as much as 15 percent over sticker. The leasing company has a contract with you and their decision to approve a lease for you is based on your credit worthiness, not the measure of the deal. As long as you pay the lease as agreed, price means nothing to them. Remember that the dealer is just as much a customer to the leasing company as you are. And to make matters even worse, the laws in most states *do not* require that leasing companies or car dealers disclose the capitalized cost on the lease contract. Consequently, once car dealers realized the implications of this little money-maker, they immediately set up separate leasing departments within their dealerships and have concentrated on converting car *buyers* to car *lessees*. Bear in mind that when you lease a car, unless you have a sophisticated financial calculator, there is virtually no way to find out what the capitalized cost is, and it can dramatically affect your ultimate cost.

The residual value or future value of a leased car plays a significant part in the monthly rental charge. The higher the projected future value, the lower the monthly payment. One of your main goals when working a lease deal is to find a car you like with the highest residual value.

To illustrate the importance of residual values, let us evaluate some real-life numbers. We will use four different vehicles of about the same capitalized cost. Assume that each is a 48-month lease and that the leasing company is charging a rate of 13 percent. Study the chart below. It clearly proves the significance of leasing a car with a high future value.

Capitalized Cost	Residual	Monthly Payment
Car A $15,345	51 percent	$286.50
Car B 15,442	47 percent	298.19
Car C 15,115	41 percent	306.38
Car D 15,559	35 percent	330.31

With the assumption that all four of the cars in the above chart are comparably equipped, and each of them would be acceptable to you, over a 48-month lease term, Car A will cost you $2102.88 *less* than Car D—a very significant example of how residual values affect monthly payments.

Another factor to consider is that different leasing companies set different residual values on the same cars. Consequently, once you have decided on a specific car, talk to the dealer and look in your phone book to find the leasing company offering the highest residual value on the car you are interested in leasing.

Unfortunately, just like the capitalized cost, lease factors or interest rates do not appear on most lease contracts. However, most banks and independent leasing companies are fairly competitive. But don't take anything for granted. Remember that *unlike* buying a car, the only real barometer for a lease is the *payment*.

The number of miles you drive during the term of a lease can also affect the monthly payment. Typically, a lease company will allow you 15,000 miles per year without any penalties or additional charges. If you drive excessive miles, the leasing company can charge a premium using two different methods. The first and least expensive way for you is for them to reduce the residual value by charging about ten cents per mile for the excessive miles. Second, they can charge you a fee at the end of the lease term based on the excessive miles at a rate of fifteen to twenty cents a mile. This could be quite costly if you went over your limit by a lot of miles. Better to pay the penalty up front, included in the payments, than to get hit with a lump-sum charge at the end of the lease term.

Leasing is relatively simple if you understand how it works. A lease allows you to drive more car for a lower monthly payment than a purchase would, all things being equal. Another thing to consider is that with most leases, there is virtually no money down. Most leasing companies require only the first payment and a relatively small refundable security deposit at lease inception.

Here are several other things to consider. You must maintain a leased car the same way you would your own car. You are responsible for maintenance, repairs—both mechanical and physical—insurance coverage, and the yearly license fees (in most states). Excessive wear and tear at lease termination may make you liable for repairs. You must read your lease contract thoroughly to fully understand exactly what your obligations are.

Sales tax can have different guidelines on a lease than on a purchase, depending upon which state you live in. Sometimes it is computed on the selling price (to the leasing company), sometimes on the total of monthly payments. You may be required to pay the sales tax up front, at the lease inception, or include the tax in your monthly payments. The sales tax laws change occasionally in most states, so you will have to consult the department of taxation in your state for a clear answer.

Most leasing companies are more rigid on the credit qualifications of lease customers. In most cases you must meet strict income requirements as well as debt-to-income ratios. Any derogatory credit will almost certainly exclude you as a lease candidate. Remember that when you lease a car, there is virtually no down payment. The leasing company's risk is higher, so they are a little more selective about their approvals.

In conclusion, the decision to lease or purchase is a very personal one. Everyone's motivation is different. If you feel that a lease is for you, considering all of the above information, then remember one important point. The monthly payment is the true measure of a good deal. As you shop around and compare various models, the bottom line is to lease the most car for the least amount of money, because at lease end you will hand the car keys to the leasing company, and that will be that.

I strongly suggest that should you decide to lease a car, it would be prudent to invest eighty dollars in a Hewlett-Packard 12-C financial calculator. This small investment can save you hundreds—even thousands—of dollars. It will enable you to determine capitalized cost with a very simple calculation. It will also enable you to solve for any of the *five* variable factors (term, interest rate, capitalized cost, payment and residual value) by registering any *four* of them. For those of you who do not wish to invest in a financial calculator, included in the appendix on page 159 is a simple math formula which will help you accurately calculate a lease payment within about 2 percent.

13

Warranty Service and Maintenance

Now that you have your sparkling new car, it is important for you to maintain it as affordably as possible, without compromising the quality of service. There always seems to be a little confusion in trying to distinguish between warranty service and maintenance. Let me explain. Warranty service is any repair or adjustment made on your car during the factory warranty period, at no charge to the owner. These repairs *have* to be made by an authorized dealer. Any defect in the car, any squeaks or rattles, or most mechanical failures are considered warranty repairs.

Maintenance, on the other hand, is a schedule of services that the manufacturer requires the owner to perform to ensure that the factory warranty doesn't become void. There are situations when the manufacturer will deny a warranty claim that typically would be a covered repair because the owner did not fulfill the maintenance requirements. Oil and filter changes, tune-ups, tire rotations, valve adjustments—all of these factory recommended services are clearly listed in the maintenance book you received when you took delivery of your new car. Read the factory maintenance schedule carefully and follow the recommended services religiously.

Many people use the dealer for warranty service but maintain their car at a less expensive independent service facility. The argument of whether or not to maintain your car at an authorized dealer, as opposed to a less expensive service center, is quite controversial. Maintaining your car is a serious business. It is my strong opinion that all services, whether major or minor, should be performed by an authorized dealer. It is a common misconception that one does not need to service a new car at the dealer where it was purchased. Although this is true in concept, it would be most beneficial for you to service your car where you bought it, even if it's a little inconvenient. If it is impossible for you to service your car where you purchased it, I strongly suggest that you select an alternative *authorized* dealer.

Most factory-recommended maintenance schedules suggest major services at 5,000- to 7,000-mile intervals, and lube, oil and filter changes at 2,500- to 3,500-mile intervals. There are several reasons why all services should be performed by an authorized dealer, particularly the selling dealer.

First of all—and most important—there are a number of situations when a major mechanical failure that occurs *after* the factory warranty has expired is actually paid for by the manufacturer. This is called a goodwill repair. People who maintain good service records and prove to the manufacturer's service representatives that all services have been performed by an authorized dealer have a very good case when attempting to have a repair paid for by the factory, even when the warranty is expired. I see this kind of goodwill repair day in and day out. The dealer and service manager will really go to bat for you if you have a serious mechanical problem after your warranty expires, *if* you are a loyal service customer.

Second, the automobile of the Nineties is extremely complex and sophisticated. The service technicians of today are unlike the mechanics of yesterday. The amount of training and experience necessary to maintain or repair a car today is extensive. Most technicians attend regularly scheduled, factory-sponsored instruction classes that keep them up to date on all phases of new repair technology and maintenance. I wouldn't want anyone other than a certified, factory-trained technician working on my car. Open the hood on your car and take a close look at the engine compartment. It's pretty overwhelming under there. Authorized dealers usually cost a little more than many independent service centers, but all in all, I believe that it is a worthy investment.

Third, if you service regularly at an authorized dealer, you will find that service advisers, service managers, and the technicians themselves will jump through hoops to help you any way they can. They will do things for you above and beyond the call of duty if you are recognized as a regular customer. I know that you cannot put a dollar value on this sort of thing, but believe me, over the years a strong relationship with a reputable, authorized dealer is priceless. I know that I am repeating myself, but it is worth reinforcing. If the day ever comes that you encounter a serious mechanical failure and your factory warranty has expired, a history of authorized service, along with the lobbying power of a service manager and the dealer himself, can often result in a no-charge repair that normally would have been out-of-pocket money.

Another thing to consider is the fact that getting your minor services such as oil and filter changes performed at a quick-change center only saves you a few dollars. It really isn't worth the nominal savings.

Finally, the car dealer wants to keep you as a customer. He wants to sell you and your family car after car for years to come. He has to treat you better than an independent service center to keep you happy and to ensure that you will continue to be a repeat customer.

If you achieved the best price on your new car from a dealer that is located a distance from your house or place of business, you may find that an alternative service facility would be more practical. There is a slight problem, though. Unfortunately, even though car dealers are smart businesspeople, sometimes they might not want to service a car that was purchased elsewhere. Some dealers cop an attitude when asked to service a car that they didn't sell, particularly if all they are asked to do is the warranty repairs. If you are faced with this situation, you should ask the dealer right up front if he has any problem servicing your car, even though you bought it someplace else. If he is a wise businessman, he will welcome you with open arms and hope that next time he can steal you away from the dealer that sold you the car.

Please do not have your car serviced at an independent service center just because you think you're saving a little money. No matter how much you think you save, an authorized dealer is the best place to get the best service. If it costs a little more, it will be well worth it. If the time ever comes that you need a friend, you would be hard pressed to find a better ally than a car dealer where you have been a loyal service customer.

I'd like to remind you to read your maintenance schedule and your owner's manual carefully. There are a number of things that you should know about your new car. Although you may find this a tedious chore, it is truly worth the effort.

Summary and Recap

1. Be sure to read this *entire* book twice.
2. The car dealer of today is a cunning, well-educated businessman.
3. Never, ever shop for a "best price."
4. Always negotiate a deal approximately five to eight days before month's end.
5. Disregard any and all new car advertising and sale events offering "sale prices."
6. Do not purchase rustproof, paint sealant, or fabric guard.
7. Do not include dealer-installed accessories in the negotiating process.
8. Check out local automotive specialty shops for lower prices on accessories.
9. Be sure that you *fully* understand the guidelines of any rebates or interest-rate incentive programs.
10. Do not select a specific car solely because of an incentive program.
11. Allow the dealer to *assume* that you will utilize his financing.
12. Finance through the dealer only if he

can meet or beat the lowest rate or if you have poor credit.

13. Consider an extended warranty *only* if it is offered by the manufacturer, it is clearly a bumper-to-bumper warranty, and the negotiated price is reasonably priced.

14. Do not commit to buying a car until you are totally prepared (avoid "if-I-could-would-you" situations).

15. Establish the ACV of your trade-in by shopping your car to used-car and new-car dealers.

16. Refer to NADA used-car guide only if you cannot get a viable offer on your trade-in. Refer to *loan value* to determine the ACV, and remember to account for noted options.

17. Clean your trade-in, inside and out, and repair minor mechanical problems.

18. Do not confuse GTA and ACV.

19. Carefully scrutinize your personal budget using the Budget and Payment Worksheet.

20. Before selecting a specific car, consider functions and features, frequency of repairs, and resale value.

21. When visiting dealers for information, make your intentions clear.

22. Leave the dealership immediately if salesman or sale manager tries to pressure you.

23. *Do not* talk price while gathering information.

24. Remain calm and reserved during the negotiating process.
25. Do not, under any circumstances, waver from your game plan or budget.
26. Do not allow dealer to switch your cash difference offer to a payment per month offer.
27. Stand your ground, even if you have to leave the dealership to get your price.
28. If you lease a car, the monthly payment is the only measure of a good deal, all things being equal.
29. Strongly consider purchasing a financial calculator if you decide to lease.
30. Do not service your new car at a non-dealer shop.
31. Drop me a line. I would love to hear from you!

A Final Word

I sincerely hope that this book has shed new light on buying a new car. I can assure you that if you trust my judgment and follow my advice, you will be rewarded for years to come. I am confident that your future car-buying experiences, from this day forward, will be unlike those of the past. I wish you the best of luck.

I invite all of my readers to drop me a line to tell me about their buying experiences. I will make every effort to respond to any questions you might have. Please include a self-addressed stamped envelope. Send all correspondence to:

DANIEL M. ANNECHINO, AUTHOR
c/o NAL/DUTTON
375 HUDSON ST.
NEW YORK, N.Y. 10014-3657

APPENDIX

Retail Buyer's Order

Buyer _____ Date _____ 19 ____ Birth Date _____ 19

Street _____ Home Phone _____ Business Phone _____

City _____ State _____ Zip _____ Salesman _____

Item	Amount	Item	Amount
Base Price	$11,175.00	Total List Price	$12,500.00
Freight	350.00	Less Trade-in or Discount	– 3,289.19
Accessories:		Net Taxable Price	$ 9,211.81
AM/FM Radio	275.00	Sales Tax	+ 644.76
Auto Trans.	700.00	License, Title, Registration	N/A*
		Inspection Fee	N/A*
		Total Cash Price	$ 9,856.57
		Less Deposit	– 500.00
		Plus Balance on Trade	+ 750.00
		Cash on Delivery	– 700.00
Total List Price	$12,500.00	Balance to Finance	$ 9,306.57

This is the top half of a typical retail buyer's order. Compare this to the Budget and Payment Worksheet that follows. It is imperative that you understand the significance of each of these numbers. The net taxable price is the same as the maximum cash difference and the balance to finance is the same as the maximum amount you can finance. Note the trade-in figure of $3,289.19. This number is a combination of the $2,000 ACV and a $1,289.19 discount. Study this retail buyer's order and the Budget and Payment Worksheet until the mathematics become totally clear.

*I have not used arbitrary numbers for license and title fees. It is assumed that these incidental amounts will be paid out of pocket.

Budget and Payment Worksheet
(SAMPLE)

[1] Maximum Monthly Payment: $ 250.00
[2] Desired Term in Months: 48
[3] Projected Interest Rate: 12.5%
[4] Monthly Payment Factor: .02658
[5] Sales Tax Rate: 7%
[6] Maximum Down Payment: 1,200.00
[7] ACV of Trade-in: 2,000.00
[8] Closeout on Trade-in: 750.00

Calculation

(1) **Maximum Amount to Finance:**
    $250 DIVIDED BY .02658   =   **$ 9,405.57***

(2) **Add to Down Payment:**
    $9,405.57 + 1,200.00   =   **10,605.57**

(3) **Subtract Closeout:**
    $10,605.57 − 750.00   =   **9,855.57**

(4) **Remove Sales Tax:**
    $9,855.57 divided by 1.07 =   **9,210.81****

(5) **Add to ACV of Trade-In:**
    $9,210.81 + 2,000.00   =   **11,210.81*****

(6) **Subtract Maximum Dealer Profit:**
    $11,210.81 − 500   =   **10,710.81******

* This is the maximum amount you can finance to achieve your desired monthly payment of $250.

** This is the maximum cash difference you can pay for any car.

*** This is the total equity available for a new car.

**** This is the maximum dealer invoice of any car considered to maintain your payment objective.

Note: See Chapter Nine, Putting It All Together, page 93, for a complete explanation of this worksheet.

Budget and Payment Worksheet

[1] Maximum Monthly Payment: ___350___
[2] Desired Term in Months: ___60___
[3] Projected Interest Rate: ___8.50___
[4] Monthly Payment Factor: ___.02052___
[5] Sales Tax Rate: ___8.25___
[6] Maximum Down Payment: ___1,000___
[7] ACV of Trade-in: ___—___
[8] Closeout on Trade-in: ___—___

Calculation

(1) **Maximum Amount to Finance:** = ___18,311___ *

(2) **Add to Down Payment:** = ___19,894___

(3) **Subtract Closeout:** = ___—___

(4) **Remove Sales Tax:** = ___1434___ / ___1874___ **

(5) **Add to ACV of Trade-In:** = ___—___ ***

(6) **Subtract Maximum Dealer Profit:** = ___17372___ / ___500___ ****

* This is the maximum amount you can finance to achieve your desired monthly payment of _____ .

** This is the maximum cash difference you can pay for any car.

*** This is the total equity available for a new car.

**** This is the maximum dealer invoice of any car considered to maintain your payment objective.

Note: See Chapter Nine, Putting It All Together, page 93, for a complete explanation of this worksheet.

Monthly Payment Factor Chart

Term in Months

APR%	24	36	48	60	66
8.50	.04546	.03157	.02465	.02052	.01902
8.75	.04557	.03168	.02477	.02064	.01902
9.00	.04568	.03180	.02489	.02076	.01927
9.25	.04580	.03192	.02500	.02088	.01939
9.50	.04591	.03203	.02512	.02100	.01951
9.75	.04603	.03215	.02524	.02112	.01964
10.00	.04614	.03227	.02536	.02125	.01976
10.25	.04626	.03238	.02548	.02137	.01988
10.50	.04638	.03250	.02560	.02149	.02001
10.75	.04649	.03262	.02572	.02162	.02014
11.00	.04661	.03274	.02585	.02174	.02026
11.25	.04672	.03286	.02597	.02187	.02039
11.50	.04684	.03298	.02609	.02199	.02052
11.75	.04696	.03310	.02621	.02212	.02064
12.00	.04707	.03321	.02633	.02224	.02077
12.25	.04719	.03333	.02646	.02237	.02090